Accounting, Finance, Sustainability, Governance & Fraud: Theory and Application

Series Editor

Kıymet Tunca Çalıyurt, Iktisadi ve Idari Bilimler Fakultes, Trakya University Balkan Yerleskesi, Edirne, Turkey

This series acts as a forum for book publications on current research arising from debates about key topics that have emerged from global economic crises during the past several years. The importance of governance and the will to deal with corruption, fraud, and bad practice, are themes featured in volumes published in the series. These topics are not only of concern to businesses and their investors, but also to governments and supranational organizations, such as the United Nations and the European Union. Accounting, Finance, Sustainability, Governance & Fraud: Theory and Application takes on a distinctive perspective to explore crucial issues that currently have little or no coverage. Thus the series integrates both theoretical developments and practical experiences to feature themes that are topical, or are deemed to become topical within a short time. The series welcomes interdisciplinary research covering the topics of accounting, auditing, governance, and fraud.

More information about this series at http://www.springer.com/series/13615

Kıymet Tunca Çalıyurt
Editor

Integrity, Transparency and Corruption in Healthcare & Research on Health, Volume I

Springer

Editor
Kıymet Tunca Çalıyurt
Trakya University
Edirne, Turkey

ISSN 2509-7873 ISSN 2509-7881 (electronic)
Accounting, Finance, Sustainability, Governance & Fraud: Theory and Application
ISBN 978-981-15-1423-4 ISBN 978-981-15-1424-1 (eBook)
https://doi.org/10.1007/978-981-15-1424-1

This Springer imprint is published by the registered company Springer Nature Singapore Pte Ltd.
The registered company address is: 152 Beach Road, #21-01/04 Gateway East, Singapore 189721, Singapore

Acknowledgements

As book series editor, I have thanks to authors and also two distinguished professors on health science.

My first thank goes to distiquished Prof. Taryn Vian from the University of San Francisco. I was reading her articles before we met in person but after visiting her office at Global Health, Boston University in 2014, my vision has widened on ethical issues in the health sector.

My second thank goes to Prof. Jillian Kohler from Toronto University. We decided to write this book after a panel titled "*Symposium: Combating Corruption in Health Care and Pharmaceuticals*" organized by Prof. Jillian Kohler on November 28, 2016. She is acting as president for the *World Health Organization Collaborating Centre for Governance, Accountability, and Transparency in the Pharmaceutical Sector*[1] at Toronto University.

IGonGFESR

International Group on Governance Fraud Ethics & Social Responsibility (IGonGFESR)

IConGFESR

International Conference on Governance Fraud Ethics & Social Responsibility (IConGFESR)

[1] [1] https://pharmacy.utoronto.ca/wp-content/uploads/sites/default/files/upload/who_cc/Official%20agenda%20schedule%20ver20161030a.pdf.

If you read this book, we owe these two distinguished professors and authors around the world.

Kıymet Tunca Çalıyurt

Introduction Chapter: How to Develop Governance in Health Companies

Kıymet Tunca Çalıyurt

> *"After spending two weeks recovering from heart and lung problems, a Florida man in his early nineties was released from an Orlondo-area hospital. His condition was too unstable for him to return home, so his doctor arranged for a short stay at a nursing facility. On his discharge from the hospital, the patient told staff members that someone was available to drive him to his new quarters. But the hospital staff insisted he makes the short trip by ambulance, assuring him that "Medicare will pay for it." The ambulance ride went smoothly, and the patient, fully conscious and aware of what was happening during the trip, arrived safely at the nursing center. Several weeks later, he received a 627USD invoice from the ambulance company. The bill for the eight-mile journey included charges for services such as the availability and actual use of oxygen (two separate charges) as well as something called an -OSHA sanitary procedure-. The oxygen, for which he was charged, was never provided, and contrary to what he was told by hospital personnel, Medicare balked at paying for the ambulance service. This case is person one because an elderly man I m describing is my father." (Leap, 2011) [1]*

This story taken from the book written by Terry L. Leap is an ordinary case that we can face within developed and also developing countries every day. When we talk about the accountable, sustainable, and ethical health sector, we also should discuss and search on the following issues. Founding Compliance Department, which controls audit and report all following duties in a health institution, is a valuable recommendation.

- Fraud examination and reporting in the health sector,
- Internal control procedures and reporting in the health sector,
- Corporate Governance and reporting in the health sector,
- Sustainability and reporting in the health sector,
- Business ethics in the health sector,
- Taxation, taxation planning in the health sector. [2]

Here, we should determine the list of healthcare institutions which means a public or nonprofit organization within this state that provides health care and related services, including but not limited to the provision of inpatient and outpatient care, diagnostic or therapeutic services, laboratory services, medicinal drugs, nursing care, assisted living, elderly care and housing, including retirement communities, and equipment used or useful for the provision of health care and related services. (Oregon, 2019) These institutions should follow the following issues other than financial statement regulations to be sustainable, reliable, comparable, ethical institutions.

Table 1 Good Governance Checklist for Health Institutions

Good governance issue for health care institutions	Issues to Discuss
Sustainability (GRI)[2] (UNGC 10 Principles)[3,4]	Implementation Sustainability
	Recruiting Sustainability Staff
	Sustainability Reporting
	Sustainability Committee
	Sustainability Certification
	Sustainability Office
Corporate Governance (OECD)	Corporate Governance Committee
	Corporate Governance Reporting
	Recruiting Corporate Governance Staff

(continued)

[2]Gal (2018) has stated that firms have increasingly used diverse channels to disclose information about their socially responsible activities to different stakeholder groups. For disclosures to regulators, the format is usually quite restricted, and firms have received some specific guidance on the method, the format, and the content of these communications. (Security and Exchange Commission, 2010) However, in making these disclosures to other groups there are a multitude of channels and formats available. One of the more formal avenues of disclosure is the reports that use the Global Reporting Initiative (GRI) format. (GRI 2006, 2011).

[3]The UN Global Compact initiative is leadership platform that aimed at encouraging business for the establishment, performing, and publishing of responsible and sustainable corporate procedures and implementations (UNGC, 2014) Global Compact seeks to promote companies to be member of this voluntary organization by 10 principles. The principles are classified into four cornerstones: human rights, labor, environment, and anti-corruption. (Çalışkan and Esen, 2018).

[4]UNGC, The Ten Principles of the United Nations Global Compact, https://www.globalcompact.ca/about/ungc-10-principles/.

Table 1 (continued)

Good governance issue for health care institutions	Issues to Discuss
Internal Audit Internal Control—Integrated Framework (COSO)[5,6,7]	Internal Control Procedures
	Cybersecurity
	Internal Control Reporting
	Recruiting Internal Audit Staff
	Certification for Internal Audit Staff
Sustainable Taxation[8] (National Code and International OECD Models)	Tax Code Implementation
	Tax Code Planning
	Tax Declaration
	Recruiting tax expert who knows tax exemptions on health sector
Corporate Reputation[9]	Evaluating reputation

(continued)

[5]Originally formed in 1985, COSO (Committee of Sponsoring Organizations of the Treadway Commission) is a joint initiative dedicated to improving organizational performance and governance through effective internal control, enterprise risk management, and fraud deterrence. COSO is jointly sponsored by the American Accounting Association (AAA), the American Institute of CPAs (AICPA), Financial Executives International (FEI), The Institute of Internal Auditors (IIA), and the Institute of Management Accountants (IMA).

[6](In USA) General Accountability Office (GAO)'s *Standards for Internal Control in the Federal Government* provides that internal control be designed to ensure that all transactions and other significant events be clearly documented and the documentation be readily available for examination. (GAO/AIMD-00-21.3.1.) To help combat fraud and abuse in healthcare programs including Medicare and Medicaid, Congress enacted the Health Insurance Portability and Accountability Act of 1996. GAO makes 11 recommendations to Department of Health and Human Services (HHS) and Department of Justice to revise or develop written procedures that include documentation and monitoring controls for Health Care Fraud and Abuse Control Program (HCFAC) activities and reporting. (Townsend, 2013).

[7]Dawson, S., Babinchak, J. (2019) COSO Issues Guidance for Healthcare Providers Collaboration with Crowe, CommonSpirit Health Aims to Strengthen Industry Governance and Internal Control, https://www.coso.org/Documents/COSO-Issues-Guidance-for-Healthcare-Providers.pdf.

[8]Due to market dynamics, taxpayers and authorities interact continuously and their interests are often interwined. Consequently, sustainability-driven goals may appear on their agendas, as they serve the interests of both. It goes without saying that corporate and individual taxpayers can display easier a sustainable tax behaviour if authorities set up a tax framework to achieve green policy objectives. The characteristics of this framework (i.e., tax incentives and tax penalties) are captured economies (KPMG, 2013) Ranging on a scale from 0 to 50, the index is meant to inform corporate taxpayers of the facilities and sanctions implemented by governments in order to promote sustainable business practices. (Batreance and et al., 2018).

[9]Caliskan and Esen have stated that there are numerous methods and tools to measure corporate reputation; however, there is no consensus about it. Some scholars as Siltaoja (2006) supports that there is no right set of criteria because different evaluators use different concepts. Several scholars have used some methods to measure corporate reputation. Reputation quotient which was devised by Haris Interactive is most common measurement. (Caliskan and Eser, 2018).

Table 1 (continued)

Good governance issue for health care institutions	Issues to Discuss
	Data collection
External Auditing (IFAC, IAASB)	International Auditing Standards
	Engagement with Audit company
	External audit committee
	External audit reporting
Environmental Reporting	Environmental regulation
	Environmental regulation implementation
	Environmental reporting
	Environmental data collection
	Reducing plastic use
Fraud Examination (ACFE)	Hiring Certified Fraud Examiner
	Fraud Prevention Procedures
	Hotline
	Fraud Examination Reporting
Business Ethics	Conduct Ethical Codes
	Reporting to institutions on ethics
Corporate Social Responsibility (CSR)	Recruiting CSR staff
	CSR reporting
	Tax exemption on CSR in health companies
	Intergrating reporting

Especially with the development of technology, the health sector has shown significant progress. This development in the health sector has left all the elements in the concept of health obliged to act in a meaningful, harmonious, holistic, and systematic manner. In this case, the health system that has emerged from the first human being has emerged—the most significant component of the health system in the hospitals. Hospitals are the backbone of the health system. For hospitals to be effective in the health system, all subsystems it contains should work with full efficiency. These systems have essential functions for the operation of the hospital. Hospitals, the most significant part of health institutions, they are for-profit or nonprofit, vulnerable to corruption. In the U.S., healthcare fraud has been estimated to cost $60 billion per year, or 3% of total health care expenditures—much of it in the hospital sector. Hospitals account for 50% or more of healthcare spending in many countries. Fraud and corruption in hospitals negatively affect access and quality, as public servants make off with resources that could have been used to reduce out-of-pocket expenditures for patients or improve needed services. (Musau, 2008) Association of Certified Fraud Examiner's Report 2018 has revealed that the most crucial fraud in the health sector is Corruption, with a rate of 36%. In the second line, we see "billing" with a rate of 26% (Fig. 1).
(Report to the Nations, ACFE, 2018)

INDUSTRY	Cases	Billing	Cash larceny	Cash on hand	Check and payment tampering	Corruption	Expense reimbursements	Financial statement fraud	Noncash	Payroll	Register disbursements	Skimming
Banking and financial services	338	11%	14%	23%	12%	36%	7%	8%	11%	2%	3%	9%
Manufacturing	201	27%	8%	15%	12%	51%	18%	10%	28%	5%	3%	7%
Government and public administration	184	15%	11%	11%	9%	50%	11%	5%	22%	7%	2%	11%
Health care	149	26%	7%	13%	13%	36%	16%	11%	19%	17%	1%	12%

Fig. 1 What are the most common occupational fraud schemes in various industries?

As in the other departments in hospitals, some tools should be applied in order to prevent abuse in hospitals. One of these is sustainability.

Sustainability

Being sustainable and accountable for hospitals are also very important, especially for publicly held companies. As they want to have to disclose their financial and nonfinancial activities to their shareholders, following full set of financial statements should be announced to hospital's stakeholders;

1. Statement of financial position,
2. Statement of comprehensive income or an income statement,
3. Statement of changes in equity,
4. Statement of cash flows, and
5. Notes.

It is easy to prepare financial statements because as accountants we have IFRS (GAAS in the USA), and auditors have IASs (International Auditing Standards).[10,11] However for sustainable hospital management, we need to disclosure reports on sustainability. As we have millions of health institutions around the world, only 420 profit or nonprofit health and pharmaceutical companies signed the agreement and promised to follow UNGC ten principles. According to CEO's of the companies who joined to UNGC states that

- Corporate and organizational success requires stable economies and healthy, skilled, and educated workers, among other factors. Moreover, sustainable companies experience increased brand trust and investor support.
- Companies offer fresh ideas and scalable solutions to society's challenges—precisely what we need to create a better world. More than 8,000 business participants and 4,000 non-business participants in the UN Global Compact are already changing the world. They are helping alleviate extreme poverty, address labor issues, reduce environmental risks around the globe, and more.

"Transforming our world: the 2030 Agenda for Sustainable Development", which was promulgated by United Nations General Assembly in 2015 has a significant impact on the business world and its agenda. This agenda is a plan of action for people, planet, and prosperity. The Sustainable Development Goals (SDGs)[12] highlight the critical role of governance across all sectors, including health. (UN, 2015) Paschke and colleagues have stated that sustainable development goals (SDGs) highlight the critical role of governance across all sectors, including health SDG 16 concerns

(i) promoting the rule of law;
(ii) preventing corrupt practices;
(iii) developing accountable and transparent institutions;
(iv) ensuring responsive, inclusive, and participatory decision-making processes; and
(v) ensuring public access to information (Figs. 2 and 3).

Source: UNGC, 2019
https://www.unglobalcompact.org/what-is-gc/participants/search?utf8=%E2%9C%93&search%5Bkeywords%5D=&search%5Bsectors%5D%5B%5D=10&search%5Bper_page%5D=10&search%5Bsort_field%5D=&search%5Bsort_direction%5D=asc

[10]IASs are published by the International Auditing and Assurance Standards Board under the International Federation of Accountants.

[11]https://www.iaasb.org/clarity-center/clarified-standards.

[12]The UNGC has announced 17 Sustainable Development Goals and 169 targets.

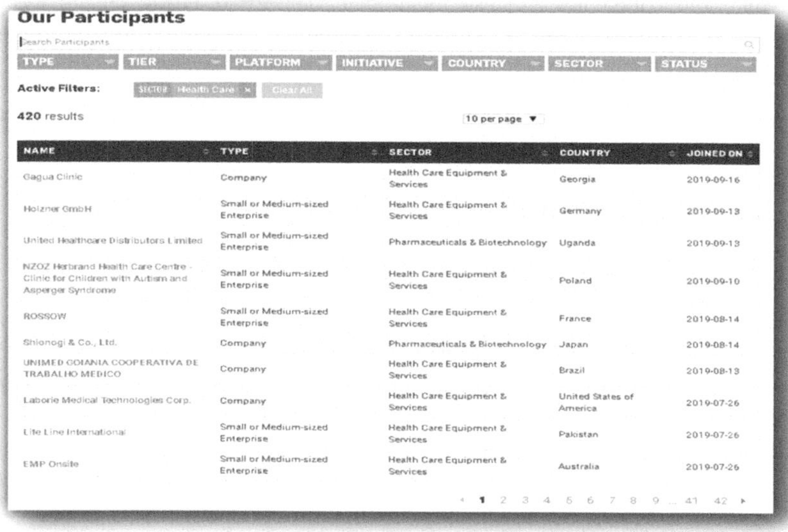

Fig. 2 Health and Pharmaceutical Profit and Nonprofit Companies Signed United Nations Global Compact (UNGC) Agreement. *Source* UNGC, 2019

Fig. 3 Benefits Of Joining UNGC

Paschke and colleagues are also mentioned "The 2010 World Health Report stated that an estimated 20–40% of potential health gains from health spending are lost through inefficiencies, such as[13]

(i) the underuse of generic medicines and the overuse of overpriced medicines, (World Health Organization; 2010)
(ii) the availability of substandard and falsified medical products;
(iii) inappropriate and ineffective prescribing; and
(iv) losses from the health system due to waste, corruption, and fraud. (Vian, 2008) (Cohen-Kohler, 2006) (Paschke and et al., 2018).

The concept of corruption in relation to global health has been defined as: "misappropriation of authority, resources, trust or power for private or institutional gain that has adverse effects on regional, local or international health systems and/or that negatively impacts individual patient and/or population health outcomes". As reported by Transparency International, the scale and scope of corruption impacting health is immense. Exact numbers are elusive, but it is estimated billions of dollars are lost annually due to corruption and fraud in a global health market estimated to be worth 10% of global gross domestic product in 2009 (Mackey, Liang, 2012).

The prevailing view is that corruption mainly affects investment and economic growth adversely. A payment of a bribe to get an investment license, for example, clearly reduces the incentive to invest (Bardhan, 1997).

In the first volume of the subtitle of book series, titled "Integrity, Transparency, Corruption in Healthcare & Research on Health Volume 1", distinguished academicians around the world have discussed the different perspective of ethics and fraud from a different perspective.

Paul D. Thacker' goal of his chapter is to discuss corporate finances influence on many areas of science originating which hired public relations firms to protect their profits from research on the harms of smoking. He stated that because corporate influence is so pervasive and often denied, policymakers must understand this history as well as the research on financial conflicts of interest to protect the public.

Fred Gifford's goal of his chapter is to examine the Health Impact Fund, a proposal aimed at solving the problem of access to pharmaceuticals for the global poor, focusing on its potential implications vis-à-vis the problem of corruption in the pharmaceutical industry. One set of questions concerns the extent to which the Health Impact Fund can be expected to help lessen or prevent the sorts of corruption that presently plague our system of pharmaceutical development. The second set of questions concerns the extent to which the sorts of structures and processes that are set up for the Health Impact Fund—making assessments of the impact of a given medicine on global disease burden—would themselves be vulnerable to corruption.

Esra Atabay and Engin Dinç' goal of their chapter is to assess whether the ISAE 3402 standard is applicable for the control and auditing of support services purchased in private healthcare organizations and to explain through a model how the

[13]https://www.unglobalcompact.org/participation/join/benefits.

ISAE 3402 standard will be implemented. To achieve the objective of the study, a survey was conducted on the managers of 25 private healthcare organizations that are included in a Private Healthcare Group with a Hospital Chain. The data obtained were analyzed by using cross-examination, single and multiple frequency analysis, and descriptive statistics. As a result of the analysis of the data obtained from the study, it was revealed that nearly half of the private hospital managers who participated in the study had heard about the ISAE 3402 standard and the vast majority believed that the assurance engagement conforming to this standard was feasible in the private healthcare industry. Moreover, it was explained via a model how the ISAE 3402 Standard would be applicable.

Mohammad Nurunnabi's goal of his chapter is to discuss on corruption in the health sector. He has stated that due to a lack of research, this chapter focuses upon corruption in health care in developing countries. Based on the extensive literature search, this study finds that corruption is significantly increasing in health care. Also, the local and international policymakers seriously address how to combat corruption in healthcare.

Joel Lexchin's goal of his chapter is to discuss the long-standing relationship between Health Canada and the pharmaceutical industry has corrupted the Canadian regulatory system. This thesis will be explored first by reviewing the history of industry–Health Canada relations and how it demonstrates an ongoing cooperative relationship whereby Health Canada has been willing to delegate regulatory responsibilities to the industry and prioritize industry's views over those of others. The chapter then looks at how clinical trials are regulated and how the desire of industry for trials to be done quickly and as inexpensively as possible means that economic values can outweigh scientific ones. The chapter then turns to the regulatory review process and how the quality of evidence that Health Canada accepts and how quickly it reviews that information impacts the efficacy and safety of drugs that reach the market. The promotion has a significant effect on how drugs are prescribed by physicians, and the chapter critically examines how Health Canada has turned over the regulation to either industry or bodies that are closely aligned with industry. Finally, the chapter looks at the deficiencies in how safety is monitored once drugs are being prescribed, sold, and used by patients.

Gabriela Arguedas-Ramírez' goal of her chapter is to discuss presenting an argument in favor of a steady-state regulation of sales representatives working for pharmaceutical companies, taking as an example the Costa Rican legislation, which can be used as part of a strategy to discourage the various incorrect practices that have created a climate prone to corruption in the relationship between prescribers and pharmaceutical companies.

Sandani Hapuhennedige, Emma Charlotte Bernsen, Jillian Clare Kohler' goal of their chapter is to discuss probing inquiry on good governance, explore the current scope of strategies that are known to be most effective, and discern any best practices that can be used to combat the issue of corruption. A literature review was conducted to specifically explore the past and current dialogue on governance, accountability, and transparency (GAT) mechanisms within international organizations (IOs). These collated findings probe thinking on crucial next steps for IOs, and the importance of establishing a stronger understanding of what mechanisms work best in practice and where. As we seek to meet the anti-corruption objectives of Sustainable Development Goal #16, building our knowledge in this area is imperative.

James Crombie's goal of his chapter is to analyse the corrupting influence of medical ghostwriting and its parallel phenomenon, guest-writing on medical and pharmaceutical research, focussing on essential differences between ghostwriting in the ordinary sense of the word—in which the ghostwriter is basically an editorial assistant recruited by the apparent author and is subject to the latter's authority—and medical ghostwriting in which the sponsor or the sponsor's agent assumes the prerogative of choosing not only the ghostwriter(s) but also the guest-writer who is invited to sign on as principal author and researcher (and who is subject to being replaced). In addition to misattribution of authorship and the apparent potential for funding bias, this type of situation involves a serious problem concerning proprietorship of and access to the raw data on which the conclusions are based. Under current laws (and the TRIPs Agreement of 1994), such data are considered to be the intellectual property of the sponsor and protected as a trade secret—a situation which is incongruent with the norms governing scientific inquiry and incompatible with concern for the well-being and safety of patients. Various strategies for preventing, reducing, and remediating the harms associated with the ghost- and guest-writing are discussed.

Adrienne Shnier's goal of her chapter is to discuss the legal concept of the "special relationship", arising in the areas of negligence and intentional torts, has been afforded little attention in products liability cases in Canadian law. The special relationship has been applied in cases wherein plaintiffs have suffered damages in the course of their relationships of reliance with defendants. Although the special relationship does not amount to a fiduciary relationship, the special relationship triggers an affirmative duty owed to the plaintiff by the defendant where the defendant makes false or misleading representations on which plaintiffs rely. Issues of limitation period expiry and discoverability are also relevant in cases where the intentional tort of fraudulent misrepresentation of a material fact relating to known risks prevent consumers from identifying their potential for causes of action before the expiry of the typical two year limitation period.

Kıymet Tunca Çalıyurt's goal in her chapter is to discuss on developing corporate governance and to prevent fraud in the procurement department. She discussed the role of the procurement department, the procedure of procurement in hospitals, and finally, fraud cases.

Roshima Said, Corina Joseph's goal of their chapter is to determine the extent of Integrity Framework information disclosure on Malaysian private hospital websites. The data collected in this study were based on a content analysis of inclusion or exclusion of integrity framework information disclosed on Malaysian private sector websites and measured using the Private Hospitals-Integrity Framework Disclosure Index (PH-IFDi). This study significantly provided useful information in helping the regulatory authorities and policymakers to strengthen their mechanisms to uphold integrity among healthcare provider. This study also has significant implications for the development of policy and guidelines for the assessment of Private Hospitals-Integrity Framework Disclosure Index (PH-IFDi).

While the others are affected by domestic and foreign developments in a variety of ways, the health sector continues its course without interruption because of "obligatory, urgent, irrecusable" reasons. Alongside this, the health sector appeals to frauds because of supervisory difficulties, it receiving a considerable percentage from the gross national product and the distribution of this percentage being done via public that is not institutionalized, and "fraud and non-budgeted costs after unethical behaviour" emerged as an important problem in the sector. (Çalıyurt, 2018, 42).

We shared researches by distinguished academicians from Canada, Turkey, Costa Rica, Malaysia and Turkey. As a result, we will be happy to receive contributions and critics from our readers.

References

Bardhan, P. (1997). Corruption and development: A review of issues. *Journal of Economic Literature,35*(3), 1320–1346.

Batreancea, L. *Sustainable tax behaviour on future and current emerging markets*, Chapter 9, Book Series: Accounting, Finance, Sustainability, Governance & Fraud: Theory and Application, In K. Caliyurt & R. Said (Eds.). Singapore: Springer Nature.

Cohen-Kohler, J. (2006). Pharmaceuticals and corruption: A risk assessment. In J. Kotalik, D. Rodriguez (Eds.), *Global corruption report 2006.* Transparency International (pp. 77–85). London: Pluto Press.

Çalıyurt, K. (2018). Ethics in health sector: Global approach. In J. Bian, K. Çalıyurt (Eds.), *Regulations and applications of ethics in business practice.* Accounting, Finance, Sustainability, Governance & Fraud: Theory and Application. Singapore: Springer.

Çalıyurt, K., Agun, C., & Pamukcu, A. (2005). *Recent developments on internal auditing legislation in public institutions of Turkey: A hospital case.* Third European Academic Conference 'Internal Audit and Corporate Governance at the Cass Business School', London, England 2005.

Çalışkan, A. O., & Esen, E. (2018). *Reputation indicators of participating companies to United Nations global compact*, Chapter 3, Book Series: Accounting, Finance, Sustainability, Governance & Fraud: Theory and Application, K. Caliyurt & R. Said (Eds.). Singapore: Springer Nature.

Dawson, S., & Babinchak, J. (2019). *COSO issues guidance for healthcare providers collaboration with crowe, commonspirit health aims to strengthen industry governance and internal control.* Retrieved March 02, 2019, from https://www.coso.org/Documents/COSO-Issues-Guidance-for-Healthcare-Providers.pdf.

Gal, G. (2018). *Are sustainability disclosures fraudulent?*, Chapter 4, Book Series: Accounting, Finance, Sustainability, Governance & Fraud: Theory and Application, K. Caliyurt & R. Said R. (Eds.). Singapore: Springer Nature.

Global Reporting Initiative (2006). *Sustainability reporting guidelines.* Global Reporting Iniative, Reporting.

Leap, T. L. (2011). *The big picture*, Phantom Billing, Fake Prescriptions, and the High Cost of Medicine, Cornell University.

Mackey, T. K., & Liang, B. A. (2012). Combating healthcare corruption and fraud with improved global health governance. *BMC International Health and Human Rights*, *12*(1), 23. https://doi.org/10.1186/1472-698x-12-23

Maruthappu, M., & Keogh, B. (2015). The future of health system leadership. *Lancet, 385*(2325). Retrieved February 23, 2019, from https://doi.org/10.1016/s0140-6736(15)60990-4.

Musau, S. (2008). *Fraud in hospitals* (4 pp.). Bergen: Chr. Michelsen Institute (U4 Brief 2008:8). Retrieved March, 15, 2019, from https://www.u4.no/publications/fraud-in-hospitals.

Oregon (2019). Health care instition. Retrieved June, 5, 2019 https://www.oregonlaws.org/glossary/definition/health_care_institution.

Paschke, A., Dimancesco, D., Vian, C., Cohen-Kohler J., & Forteb, G. (2018). Increasing transparency and accountability in national pharmaceuticalsystems. Bull World Health Organ, *96*:782–791.

Securities and Exchange Commission. (2010). *Commission guidance regarding disclosure related to climate change*, final rule 17 CFR parts 211, 231, 241. Securities and Exchange Commission, Washington, DC.

Siltaoja, M. E. (2006). Value priorities as combining core factors between CSR and reputation – a qualitative study. *Journal of Business Ethics*, *68*(1), 91–111.

Townsend, L. (2013). *Health care fraud and abuse control program*, Health care issues, costs and access. New York: Nova Science Publishers.

UNGC. (2014). *How to Participate UN Global Compact*. Retrieved July 21, 2014 http://www.unglobalcompact.org/Docs/News_Events/8.1/GC_Brochure_FINAL.Pdf.

UNGC. (2019). *Health and Pharmaceutical Profit and NonProfit Companies Signed United Nations Global Compact (UNGC) Agreement.* https://www.unglobalcompact.org/what-is-gc/participants/search?utf8=%E2%9C%93&search%5Bkeywords%5D=&search%5Bsectors%5D%5B%5D=10&search%5Bper_page%5D=10&search%5Bsort_field%5D=&search%5Bsort_direction%5D=asc, Accessed on Dec 19th, 2019.

United Nations. (2015). *United nations general assembly opens on Sep 15th 2015*. https://www.un.org/africarenewal/africaga2015/united-nations-general-assembly-opens-15-september-2015 (Accessed on March 5th, 2019).

Vian, T. (2008 March). Review of corruption in the health sector: theory, methods and interventions. *Health Policy and Planning*, *23*(2), 83–94.

WHO. (2010). Health systems financing: The path to universal coverage. *The World Health Report 2010*. Geneva: World Health Organization.

Contents

Part I Transparency and Accountability in Healthcare & Pharmaceutical Sector

1 **Transparency and Conflicts in Science: History of Influence, Scandal, and Denial** 3
Paul D. Thacker

2 **The Health Impact Fund and the Problem of Corruption in the Global Pharmaceutical Sector** 27
Fred Gifford

3 **Outsourced Services in Private Healthcare Organizations and ISAE 3402 Assurance Engagement: A Research Focused on Turkey** 41
Esra Atabay and Engin Dinç

4 **Revisiting Accountability: Corruption in Health Care in Developing Countries** 65
Mohammad Nurunnabi

Part II National & International Regulation on Fraud Prevention in Healthcare & Pharmaceutical

5 **Corruption of the Canadian Drug Regulatory System** 81
Joel Lexchin

6 **Professional Ethics, Professionalization, and Regulation of Pharmaceutical Sales Representatives: Analyzing the Costa Rican Case** 105
Gabriela Arguedas-Ramírez

7 **Exploring Accountability and Transparency Within International Organizations: What Do We Know and What Do We Need to Know?** . . . 125
Sandani Hapuhennedige, Emma Charlotte Bernsen and Jillian Clare Kohler

8 **Medical Ghost- and Guest-Writing as Corrupt Practices and How to Prevent Them** . . . 141
James Crombie

Part III Fraud Schemes & Cases & Fraud Prevention in Healthcare & Pharmaceutical Sector

9 **Fraudulent Misrepresentation and Fraudulent Concealment in Products Liability in Tort Law in Canada: The Special Relationship Between Drug Companies and Consumers in the Context of the Fraudulent Misrepresentation and Fraudulent Concealment of Data** . . . 161
Adrienne Shnier

10 **Developing Governance of Procurement Department in Hospitals** . . . 217
Kıymet Tunca Çalıyurt

11 **Communicating Integrity Framework Information on Malaysian Private Hospitals' Websites** . . . 237
Roshima Said and Corina Joseph

Index . . . 249

About the Editor

Prof. Dr. Kıymet Tunca Çalıyurt CPA, CFE, graduated from the Faculty of Business Administration at Marmara University, Istanbul, Turkey. Her Masters and Ph.D. degrees are in the Accounting and Finance Programme from the Social Graduate School, Marmara University. She has worked as an auditor in Horwath Auditing Company, a manager in McDonalds and finance staff in Singapore Airlines. After vast experience in the private sector, he has started to work in academia. She is holding CFE and CPE titles. Her research interests are in accounting, auditing, fraud, social responsibility, corporate governance, finance, and business ethics, with a particular interest in aviation management, NGOs, women rights in business. She has been as visiting rescarcher in Massachusetts University Amherst Business School. She is the founder of the International Group on Governance, Fraud, Ethics and Social Responsibility (IGonGFE&SR) which was founded in 2009. In 2009, she also founded the International Women and Business Group, which organizes a global, annual conference. Kiymet has published papers, book chapters, and books both nationally and internationally on fraud, social responsibility, ethics in accounting/finance/aviation disciplines in Springer and Routledge. She is book series editor in Springer with the title Accounting, Finance, Sustainability, Governance & Fraud: Theory and Application, and book series editor in Routledge with the title Women and Sustainable Business. Some book titles: Emerging Fraud (with Sam Idowu), Corporate Governance: An International Perspective (with Sam Idowu), Women and Sustainability in Business: A Global Perspective, Sustainability and Management: An International Perspective (with Ulku Yuksel), Globalization and Social Responsibility (with David Crowther), Regulations and Applications of Ethics in Business Practice (with Dr. Jiang Bian), Ethics and Sustainability in Accounting and Finance, Volume I. She is acting as member in editorial board Journal of Financial Crime, Social Responsibility Journal, International Journal of Law and Management, Journal of Money Laundering Control. She is a regular speaker at International Economic

Crime Symposium in Jesus College, Cambridge University and member in editorial board Social Responsibility Journal, Journal of Financial Crime, International Journal of Law and Management. She is a partner of Herme Consulting in Trakya University Technopark.

List of Figures

Fig. 3.1 Example 1—To assurance services model (narrowed model). *Source* Atabay (2016: 111) . . . 50
Fig. 3.2 Example 2—To assurance services model (comphrehensive model). *Source* Atabay (2016: 112) . . . 50
Fig. 4.1 Framework for analyzing corruption (Cuervo-Cazurra, 2016, p. 36) . . . 68
Fig. 4.2 Corruption ranking of Asia Pacific in 2016 . . . 70
Fig. 4.3 Corruption ranking of Americas (North and South America) in 2016 . . . 70
Fig. 4.4 Corruption ranking of MENA (Middle East and North Africa) in 2016 . . . 71
Fig. 4.5 Corruption ranking of EU (European Union) in 2016 . . . 71
Fig. 4.6 Corruption ranking of BRICS in 2017 (Brazil, Russia, India, China, and South Africa) . . . 72
Fig. 6.1 The link between promotional techniques . . . 108
Fig. 10.1 Fraud tree . . . 222
Fig. 10.2 Five components and 17 principles of internal control. *Source* Adapted from the COSO Internal Control–Integrated Framework (Shandal & Foster, 2019) . . . 223
Fig. 10.3 The fraud risk management framework. *Source* GAO Highlights, 2018 . . . 224

List of Tables

Table 3.1 Findings related to demographic features of participants 53
Table 3.2 Findings on support service procurements of private hospitals 55
Table 3.3 Findings related to control of support services 56
Table 3.4 Findings related to control and audit of support services 57
Table 3.5 General findings on the ISAE 3402 Standard 58
Table 3.6 Findings on the reasons for managers to prefer assurance engagement service in conformity with the ISAE 3402 58
Table 3.7 Findings on the reasons for managers to avoid assurance engagement service in conformity with the ISAE 3402 59
Table 3.8 Comparison between the opinion of applicability and the demand 59
Table 3.9 Findings on the justifications reported by the participants who avoided procuring the assurance engagement in conformity with the ISAE 3402 despite their opinion that the assurance engagement is applicable 60
Table 4.1 Corruption perceptions index 2016 (High-ranked countries) [*Low level of corruption/very clean*] 66
Table 4.2 Major findings of Articles on corruption and health care 73
Table 6.1 Summary of the legal framework that regulates the medical visit in Costa Rica 115
Table 9.1 Elements of the "Special Relationship" as defined by the Supreme Court of Canada 171
Table 11.1 Total items for private hospitals-integrity framework disclosure index (PH-IFDi) 242
Table 11.2 Descriptive statistics for private hospitals-integrity framework disclosure index (PH-IFDi) 242
Table 11.3 Descriptive statistics for private hospitals-integrity framework disclosure index (PH-IFDi)—by categories 242
Table 11.4 Descriptive statistics for private hospitals-integrity framework disclosure index (PH-IFDi)—rank by themes 243

Part I
Transparency and Accountability in Healthcare & Pharmaceutical Sector

Chapter 1
Transparency and Conflicts in Science: History of Influence, Scandal, and Denial

Paul D. Thacker

Abstract Corporate finances influence many areas of science, originating with tobacco companies which hired public relations firms to protect their profits from research on the harms of smoking. Despite a large body of studies finding that money biases research, scientists and academic organizations fail to embrace the peer-reviewed research on corporate influence. In many instances, they reject the science and try to rationalize behavior, leading a cycle of scandal, followed by reform, followed by later scandal. Because corporate influence is so pervasive and often denied, policymakers must understand this history as well as the research on financial conflicts of interest to protect the public.

Keywords Transparency · Conflict · Health

1.1 Introduction

In December 1953, the CEOs of America's leading tobacco companies cast aside competitive rancor and gathered at the Plaza Hotel in New York City to confront a menace to their incredibly profitable industry. An emergent body of science published in elite medical journals cast doubt on the safety of cigarettes and threatened to destroy a half-century of corporate success. Joining them at the Plaza was John W. Hill, the president of America's top public relations firm, Hill & Knowlton. Hill would later prove a decisive savior (Brandt, 2012). Hill had closely studied Edward Bernays, whose work on propaganda in the 1920s and 1930s laid the foundation of modern public relations and defined common techniques to manipulate popular opinion (Tye, 1998). Hill understood that any traditional campaign would fail to sway society which perceived advertising as little more than corporate propaganda. Effective public relations required comprehensive off-stage management of the media. At its best, it left no fingerprints (Brandt, 2007).

P. D. Thacker (✉)
Edmond J. Safra Center for Ethics, Harvard University, Cambridge, USA
e-mail: thackerpd@gmail.com

K. T. Çalıyurt (ed.), *Integrity, Transparency and Corruption in Healthcare & Research on Health, Volume I*, Accounting, Finance, Sustainability, Governance & Fraud: Theory and Application, https://doi.org/10.1007/978-981-15-1424-1_1

Instead of ignoring or denigrating new data that found tobacco dangerous, Hill proposed the opposite: embrace science, trumpet new data, and demand more, not less research. By calling for more research, which they would then fund, tobacco companies could harness academic scientists in a battle to confront a major scientific controversy and amplify skeptical views of the relationship between tobacco and disease (Brandt, 2012). Such a scheme would let companies shroud themselves in doubt and uncertainty—core principles of the scientific process, in which every answer leads to new questions.

Hill & Knowlton's campaign for the five largest U.S. tobacco companies corrupted science and medicine for decades to follow, laying the foundation for financial conflicts of interest in science, as other industries mimicked tobacco's techniques to protect their own products from government bans and regulations—later, from consumer lawsuits (Brandt, 2012). While tactics have varied over time, the core strategy has changed little since tobacco wrote the playbook, providing a menu of techniques now employed across industries (Michaels, 2008). To position themselves as more science than the science itself, corporations hire academics as advisors or speakers, appoint them to boards, fund university research, support vanity journals, and provide academic scholars with ghostwritten manuscripts to which they can add their names and publish in peer-reviewed journals with sometimes little or no effort. These tactics create an alternative scientific realm that drowns out the voices of independent researchers and calls into question the soundness of impartial data. To further undermine impartial scientists, industries secretly supports think tanks and corporate front groups. These organizations echo and amplify company studies and experts, counter articles in the media, and launch campaigns against independent academics, often trying to get their research retracted or perceived as second-rate and untrustworthy to the public and media.

To counter corporate influence, academic and government bodies have repeatedly turned to conflict of interest policies and calls for greater transparency and financial disclosure. Philip Handler, the President of the National Academies of Science (NAS) during the early 1970s, proposed the first conflicts of interest policy which the NAS Council approved in 1971 (Jukes, 1983; Parascandola, 2007). The policy drew sharp rebukes from leading scientists who called it "insulting" and "undignified," creating a pattern that continues today (Brandt, 2012). Whenever a scandal erupts that finds companies exerting undue influence on science, calls for greater transparency and more stringent ethics requirements are countered with assertions that current rules are fine and further scrutiny is not needed.

However, a growing body of literature finds that arguments against financial conflicts of interest reforms are unsubstantiated, lacking in intellectual rigor, and ignorant of the peer-reviewed research on financial influence. Although conflicts of interest policies have become more prevalent, their content and essential requirements have evolved little since the National Academies introduced their first rules (Parascandola, 2007). In fact, the controversy over corporate control of science continues to dog the Academies to this very day. Over 40 years after introducing their first conflicts of interest policy, the Academies are once again caught in a scandal, this time after complaints that committee members preparing reports for the Academies have cozy ties to corporations (Taylor, 2017).

Investigative reporters found that nearly half the members of a 2011 Academies report on pain management had ties to companies that manufacture narcotics, including opioids (Fauber, 2014). A separate newspaper investigation discovered that the NAS staff member who selected the committee members for a report on the regulation of the biotechnology industry was simultaneously applying to work for a biotech nonprofit. Many of the committee members he chose were found to have undisclosed financial ties to biotech corporations (Strom, 2016). As this review of history will show, the Academy is not alone in confronting conflicts of interest in a cycle of denial, scandal, reform, and more denial.

1.2 Early Years

Concern over corporate influence on science is relatively modern, having emerged in the 1960s. In the early twentieth century, private foundations and research institutes funded the vast majority of scientific research in the United States. This changed after World War II when the national government began pouring increasing amounts of money into scientific programs. Physicist Paul E. Klopsteg best expressed the apprehension many scientists felt about the government controlling the research agenda. As the Associate Director for Research at the National Science Foundation in 1955, he worried that federal funding for science could allow the government to hijack the mission of universities. "Does such a vision make you uneasy?" he asked in a rhetorical fashion. "It should; for it requires scant imagination to picture therein a bureaucratic operation that would irresistibly and inevitably take a hand in the affairs of our institutions of higher learning" (Klopsteg, 1955).

The government's influence over science can be evaluated by examining budget numbers. From its first year of operations in 1952, the National Science Foundation's budget ballooned from \$3.5 to almost \$500 million in 1968. The National Institutes of Health saw equally large increases, growing from \$2.8 million in 1945 to over \$1 billion in 1967. By 1960, the government supported over 60% of research (Parascandola, 2007).

During this period, the scientific community focused on conflicts of interest that affected scientists who either worked in government or who were funded by government agencies, especially researchers in military and space science research programs ("Scientists in Government: Growing Concern Over Conflicts of Interest," 1960). Even while using the term "conflict of interest", scientists discussed the matter only within a narrow legal context. When Congress held hearings about conflicts of interest in science, they concerned scientists who were government contractors for the Atomic Energy Commission or National Aeronautics and Space Administration while also having financial interests in private research or consulting companies. Worries about government influence over science were also apparent in 1964. That year, both the American Council on Education and the American Association of University Professors developed conflicts of interest policies that only discussed research funded by the government (Parascandola, 2007).

By examining the appearance of the phrase "conflicts of interests" in the journal *Science* over the past century, we can see how the term has changed in context and meaning, reflecting researchers' concerns about the power of external forces in shaping science. In the early years, the term surfaced in the journal's pages in reference to scientists' relations to government. Over time, this shifted to incidents and discussions involving the industry. This uneasiness with industry seems to have increased with time and with strengthening kinship between universities and corporate partners (Gingras & Gosselin, 2008).

1.3 Tobacco Creates Parallel Science

After an initial meeting with tobacco company leaders in late 1953, Hill & Knowlton created a sophisticated strategy to shroud the emerging science about tobacco in skepticism (Brandt, 2012). Skeptics have always existed in science. In fact, skepticism is a fundamental value of science. But tobacco repurposed skepticism by flooding the research field with money to study the relationship between smoking and disease, and positioning the industry as scientific advocates while shaping and amplifying a public message that tobacco's potential dangers were an important scientific controversy. Historian Allan M. Brandt, of Harvard University, writes "Doubt, uncertainty, and the truism that there is more to know would become the industry's collective new mantra" (Brandt, 2012).

This Trojan Horse intrusion avoided many potential downfalls of a direct assault. Attacking researchers could backfire and be viewed as bullying; issuing statements of safety could be dismissed by a cynical public as self-serving, or worse, dishonest. But emphasizing the need for more research allowed the tobacco industry to seize the moral high ground from which they could then peer down onto emerging data, gently guiding new research, to spur a spurious debate. While pretending the goal was science, tobacco companies would repurpose research for public relations (Brandt, 2012).

Public relation firms had decades of expertise at stage-managing the media to counter information that harmed their clients. But by controlling the research agenda and the scientific process, tobacco companies could manage journalists even better than in the past. Instead of manipulating journalists to fight on their side of a public debate, companies would create the debate and then harness the media to publicize it for them (Brandt, 2012).

As part of their initial plan, tobacco companies sought experts to discredit new research that might find links between tobacco and lung cancer (Hanmer, 1954). After companies collected public statements of physicians and scientists, Hill & Knowlton then produced a compendium of experts and their quotes (Brandt, 2012). Not content with just funding individual scientists and research projects, Hill proposed creating an industry-funded research center (Inc, 1953). This call for new research broadcasted a subtle message that current data was outdated or flawed, and by partnering with academic scientists and their universities, it created the impression that the tobacco industry was committed to finding the right answers (Brandt, 2012).

"It is believed," Hill wrote, "that the word 'Research' is needed in the name to give weight and added credence to the Committee's statements" (Inc, 1958). By branding tobacco as a proponent of research, Hill made science the solution to possible government regulation. This strategy would lead to almost half a century of collusion between tobacco corporations and university researchers ("United States v Philip Morris et al.," 2004).

The Tobacco Industry Research Committee (TIRC) became central to Hill & Knowlton's strategy of co-opting academia. When TIRC was officially formed, over 400 newspapers ran an ad announcing the group with the title, "A Frank Statement to Cigarette Smokers" (Brandt, 2012). The ad noted that tobacco had been accused of causing all sorts of human diseases, yet, "One by one those charges have been abandoned for lack of evidence" (Committee, 1954). The ads then pledged that companies would fund, on behalf of consumers, new research to study tobacco's health effects:

> We accept an interest in people's health as a basic responsibility, paramount to every other consideration in our business. We believe the products we make are not injurious to health. We always have and always will cooperate closely with those whose task it is to safeguard the public health (Committee, 1954).

The Executive Director of TIRC was W.T. Hoyt, a Hill & Knowlton employee, who operated TIRC from his firm's New York office. Hoyt had no scientific experience, and before joining the PR firm, he sold advertising for the *Saturday Evening Post* (Brandt, 2012). The tobacco industry would later conclude "most of the TIRC research has been of a broad, basic nature not designed to specifically test the anti-cigarette theory" (Burling, 1963).

After retiring as CEO of Brown & Williamson, Timothy Hartnett became the first full-time chairman of TIRC. The statement announcing his appointment, reads:

> It is an obligation of the Tobacco Industry Research Committee at this time to remind the public of these essential points:
>
> 1. There is no conclusive scientific proof of a link between smoking and cancer.
> 2. Medical research points to many possible causes of cancer ….
> 3. A full evaluation of statistical studies now underway is impossible until these studies have completed, fully documented and exposed to scientific analysis through publication in accepted journals.
> 5. The millions of people who derive pleasure and satisfaction from smoking can be reassured that every scientific means will be used to get all the facts as soon as possible (Inc, 1954).

The TIRC began operating in 1954 and almost all of its $1 million budget was spent on fees to Hill & Knowlton, media ads, and administrative costs. Hill & Knowlton hand-picked TIRC's science advisory board (SAB) of academic scientists who peer-reviewed grants which had been screened previously by TIRC staff. Hill & Knowlton favored scientists who were skeptics of tobacco's ill health effects, especially skeptics who smoked (Brandt, 2012).

Instead of delving into research about tobacco's links to cancer, most of TIRC's program focused on answering basic questions about cancer in areas such as immunology, genetics, cell biology, pharmacology, and virology (Brandt, 2012). The TIRC funding of universities helped chill discourse and debate that argued tobacco might cause disease while also allowing tobacco companies the prestige of associating with academics, as few TIRC scientists took strong positions against tobacco (Brandt, 2012).

While launching TIRC, Hill & Knowlton also moved to reshape the media environment by developing a large, systematically cross-referenced library on tobacco-related issues. As one Hill & Knowlton executive explained:

> One policy that we have long followed is to let no major unwarranted attack go unanswered. And that we would make every effort to have an answer in the same day—not the next day or the next edition. This calls for knowing what is going to come out both in publications and in meetings....This takes some doing. And it takes good contacts with the science writers (Hill, 1962).

Although their positions were not grounded in substantive peer-reviewed literature, Hill & Knowlton broadcast the opinions of a small group of skeptics on cigarette science, making it appear as if their views were dominant in medical research. These skeptics allowed TIRC to quickly counter any assault against tobacco. In many cases, TIRC rebutted new findings even before they had become public (Brandt, 2012). This campaign succeeded because it hijacked science journalists' love of controversy and commitment to balance. Brandt concluded, "Given the penchant of the press for controversy and its often naive notion of balance, these appeals were remarkably successful" (Brandt, 2012).

Not satisfied with passive forms of media control like advertising and press releases, Hill & Knowlton practiced aggressive outreach to authors, editors, scientists, and other opinion makers. Personal face-to-face contacts were critical, and after every press release, TIRC would initiate a "personal contact." Hill & Knowlton systematically documented this courtship of newspapers and magazines to urge journalistic balance and fairness to the tobacco industry. During these encounters, TIRC emphasized that the tobacco industry was committed to the health of cigarette smokers and scientific research, while urging skepticism about statistical studies finding harm. Finally, TIRC presented journalists with contacts of "independent" skeptics to ensure accurate journalistic balance. In short, after creating the controversy, Hill & Knowlton then co-opted reporters to cover the debate, leading to stories that concluded tobacco science was "unresolved" (Brandt, 2012).

Despite Hill & Knowlton's behind the scenes management of TIRC to provide a veneer of scientific credibility, scientists advising TIRC balked about the board's independence and their professional credibility among peers. To calm these fears, Hill & Knowlton created the Tobacco Institute in 1958, at the behest of R. J. Reynolds.

An industry attorney later recounted that "the creation of a separate organization for public information was hit upon as a way of keeping [TIRC scientists] inviolate and untainted in [their] ivory tower while giving a new group a little more freedom of action in the public relations field" (Burling, 1963). Having protected the "science" mission of TIRC, Hill & Knowlton operated the Tobacco Institute as an effective political lobby in Washington to counter congressional hearings and potential agency regulations. As it had in advertising and media, the tobacco industry innovated new strategies with the Tobacco Institute to manipulate the regulatory and political environment (Brandt, 2012).

Hill & Knowlton's success became evident in 1961. When tobacco hired the firm in 1954, the industry sold 369 billion cigarettes. By 1961, companies sold 488 billion cigarettes, and per capita cigarette use rose from 3,344 annually to 4,025, the highest in American history (Hill, 1962).

In 1963, a *New York Times* story noted, "Surprisingly, the furor over smoking and health failed to send the industry into a slump. Instead, it sent it into an upheaval that has resulted in unforeseen growth and profits." An official with the American Cancer Society told the paper, "When the tobacco companies say they're eager to find out the truth, they want you to think the truth isn't known…. They want to be able to call it a controversy" (Lelyveld, 1963).

During this time span, scientists seemed unperturbed by the conflicts of interest that arose when tobacco-funded university research and academics allied themselves with a corporate campaign (Parascandola, 2004). When the Surgeon General established an advisory committee on smoking and health in 1963, the committee did not have a conflict of interest policy. In fact, the tobacco industry was allowed to nominate and reject committee members (Parascandola, Weed, & Dasgupta, 2006).

Although documents detailing tobacco's tactics to hijack science only became public following litigation in the 1990s, this playbook created in the 1950s remains effective and has been copied by other industries (Brownell & Warner, 2009; Oreskes, 2010). In order to disrupt scientific norms and stave off regulation, many corporations now make boilerplate claims of scientific uncertainty and lack of proof, and divert attention from product health risks by placing blame on individual responsibility (Brandt, 2012).

Before tobacco, both the public and scientific community believed that science was free of undue influence from special interests. However, tobacco repurposed science not to advance knowledge, but to undo that which was already known: cigarette smoking is dangerous. Instead of funding research to make new facts, tobacco spread money around to unmake that which was already a fact. Historian, Robert Proctor, of Stanford University has used the term "agnotology" to describe this process of constructing ignorance (Proctor, 2008).

To this day, society struggles to create policies to limit corporate influence over areas of science that advance the public interest and intersect with government regulations (Brandt, 2012). We can thank the tobacco industry for inventing our modern crisis with conflicts of interest and financial transparency in science (Thompson, 1993).

1.4 Modern Scandal

The late 1960s and early 1970s marked a period of political turmoil and social change in the United States. Confidence in government and social institutions plummeted with the Watergate scandal and a series of exposés that shined a harsh light on special interests that were manipulating Congress (Guttman, 1976). At the same time, Congress created new federal agencies with broad mandates for protecting public health, elevating the role of scientists in federal policymaking. The Environmental Protection Agency and the Occupational Safety and Health Administration, created in 1970, were charged with developing regulatory standards for a wide range of substances for which limited data existed. At the same time, the 1971 National Cancer Act brought attention to environmental factors related to cancer risk (Parascandola, 2007).

Describing this period, sociologist Sheila Jasanoff remarked that science advisors had become a "fifth branch" of the government (Jasanoff, 1998). But as medicine and science began to have a more direct impact on policy, they simultaneously came under greater public scrutiny, leading to controversies about scientific integrity. Media outlets at the time ran front-page stories about financial interests and apparent corruption regarding several issues that touched on the environment, consumer safety, and public health. Prior to this, the public was rarely confronted with evidence about the dangers of radiation, chemical pesticides, and food additives and how these substances might cause cancer. Yet, as scientists and physicians found their professions more heavily scrutinized, society also demanded that they create policies to protect public health (Parascandola, 2007).

In 1970, the National Academies faced accusations of pro-industry bias, after creating a committee to examine the health effects of airborne lead. Dupont and the Ethyl Corporation—the two companies that produced the most lead in the United States—employed 4 of the committee's 18 experts. An Academies spokesperson defended the committee, arguing that members were selected on the basis of scientific qualifications, and that they advised the Academy as scientists, not as representatives of their employers (Jukes, 1983; Parascandola, 2007).

The President of the Academies during this period was Philip Handler, a former academic who consulted for numerous food and pharmaceutical companies and served on the Board of Directors of the food corporation Squibb Beech-Nut (Jukes, 1983). Throughout his tenure, Handler continued to face criticism over his industry ties (Cohn, 1971; Parascandola, 2007).

Handler attempted to thread the needle of conflicts of interest by pointing to the Academy's obligation to work with the Department of Defense to protect the country. "[T]he question is not whether the Academy should do work for the Defense Department but how it goes about maintaining its objectivity in doing so," he argued (Nelson, 1968). Handler also advocated for more federal funding for graduate scientific education but cautioned that the "university must not become subservient to or the creature of the federal government by virtue of this financial dependency." While arguing that government and industry funding was essential for science, he seemed to sidestep the obvious dilemma that this funding might compromise scientific independence (Parascandola, 2007).

After the airborne lead committee kerfuffle, Handler proposed that new committee members disclose any potential conflicts that might arise during service for the Academy. This information would be shared among fellow committee members, not the public, and was aimed at providing information to the Academy that might be damaging if it became public through other avenues. The new conflicts of interest rules were limited to explicit financial relationships, but also considered "other conflicts," that might be perceived as creating bias (Parascandola, 2007).

Before implementing the new policy, Handler conducted an informal survey of committees and boards at NAS. Some responded that all members were in conflict, while others said scientists could not be biased. One committee member wrote, "Isn't it probably true that unless a committee member has some possibility of [conflict of interest], it is not too likely that he will be a useful committee member?" (Parascandola, 2007) In short, when scientists were prodded about the conflicts of interest and how this might bias their opinion, they inverted the problem by redefining conflicts of interest as "scientific expertise."

In August 1971, the Academy approved a one-page letter, titled "On Potential Sources of Bias," to be filled out by potential advisory committee members. The letter noted that NAS committees were, to an "ever-increasing extent," being asked to consider issues of "public interest or policy," thus frequently requiring conclusions that rested on "value judgments" as well as data. Even when committee members are acting without bias, the letter stated, such charges can impugn committee reports and conclusions. Thus, individual members were asked to state "which [factors], in his judgment, others may deem prejudicial" (Parascandola, 2007).

Many committee members viewed the statement as an accusation or challenge to their integrity, with some calling it "insulting" and "undignified." Federal laws required government advisors to disclose financial conflicts such as grants or stocks, but the Academy's statement delved into other sources of potential bias such as prior comments and membership in organizations (Parascandola, 2007).

Still, concern about the Academy's integrity reprised the following year when its Food Protection Committee was accused of pro-industry bias and downplaying the cancer risks of food chemicals. Food companies partly funded the committee which included academics, who consulted for the food industry (Gillette, 1972). Worries about industry influence were further inflamed in 1975, when Ralph Nader funded a former journalist for *Science*, Philip Boffey, to investigate the Academy's ties to industry and how corporate financial support may have influenced their reports (Boffey, 1975).

Nonetheless, the Academy's 1971 statement was a pioneering policy in conflicts of interest and the precursor to the Academy's current practices (Academies, 2003). But a new element would enter the picture in 1980 when Congress passed the Bayh–Dole Act. This law allowed universities to own inventions created by professors with government funding and encouraged corporate collaborations to develop new products and bring them to market (Thursby & Thursby, 2003). Within a year, many top academic centers and their faculty had signed lucrative licensing deals with pharmaceutical and biotechnology companies, dividing academics at American universities over uneasiness about scientific integrity and academic freedom (Culliton, 1982).

1.5 Current Evidence and Primacy of Pharmaceutical Companies

In the early 1900s, the American Association of University Professors published a declaration of principles to guide academic life. In retrospect, this declaration seems quaint:

> All true universities, whether public or private, are public trusts designed to advance knowledge by safeguarding the free inquiry of impartial teachers and scholars. Their independence is essential because the university provides knowledge not only to its students but also to the public agency in need of expert guidance and the general society in need of greater knowledge; and… these latter clients have a stake in disinterested professional opinion, stated without fear or favor, which the institution is morally required to respect (Professors, 1915).

Current university practices resemble these principles about closely as modern sexual behavior smacks of the prim morality of the Victorian era. Just as the 1960s sexual revolution altered sexual behavior, tobacco transformed university practices by blurring the boundaries between corporate public relations and academic research (Brandt, 2012). These changes have been most profound in medicine, where academic partnerships with the biotechnology industry have created both cures for several diseases and a pandemic of financial conflicts of interest (Press, 2000; Weatherall, 2000). In effect, the pharmaceutical industry has repurposed tobacco's campaign by co-opting academics to sell drugs. These financial conflicts of interest in academic biomedical research entered the public debate in the early 1980s, following a series of scientific misconduct scandals. In some cases, investigations revealed that faculty members fabricated or falsified data for products in which they had a financial interest (Korn, 2000).

By then, two important laws helped bind academics to the biotech industry. In 1980, Congress passed the Steven Wydler Technology Innovation Act and Bayh–Dole Act. The Wydler Act pushed federal agencies to transfer technologies they helped invent to the private sector, leading many universities to create technology transfer offices. The Bayh–Dole Act allowed small businesses to patent inventions created with federal grants, allowing universities to license products their faculty created. Both laws aimed to leverage federal agencies and funding to bring lifesaving products to the public. However, the laws also pushed academics into a further alliance with industry (Witt & Gostin, 1994).

As the distinction between academic research and industry marketing continued eroding, the *New England Journal of Medicine* announced the first formal conflict of interest policy for any major science journal in 1984 (Relman, 1984). In an editorial, the NEJM's editor laid out concerns that required this new policy:

> Now, it is not only possible for medical investigators to have their research subsidized by businesses whose products they are studying, or act as paid consultants for them, but they are sometimes also principals in those businesses or hold equity interest in them. Entrepreneurialism is rampant in medicine today. Any new research development that has or might have commercial application attracts attention from established corporations or venture capitalists. Reports of such developments released at press conferences, presented at scientific meetings, or published in journals may cause stock prices to rise abruptly and fortunes to be made almost overnight. Conversely, reports of unfavorable outcomes or serious side effects may rapidly devalue a particular stock. On more than one occasion during the past few years, the publication of an article in the Journal has been the direct cause of sharp fluctuations in stock prices (Relman, 1984).

A year later, JAMA also instituted a conflict of interest policy. However, the two leading science journals did not catch up until 1992 (Science) and 2001 (Nature). Research finds that science disciplines have always lagged behind medicine in addressing financial bias (Thacker, 2014).

For example, in 1990, Harvard Medical School instituted financial conflict of interest policies, by limiting the types of commercial relationships clinical research faculty could have and setting a ceiling on financial interests (School, 2010). This appears to the be the first attempt by a university to sharpen the distinction between academic research and corporate product development. Both the Association of American Medical Colleges and the Association of Academic Health Centers followed up that year by publishing guidance on financial conflicts of interest (Centers, 1990; Colleges, 1990).

In these same years, the National Institutes of Health proposed new rules to require that academics disclose financial interests to their institution and not consult or have equity in companies that might be affected by their research. In response, the NIH received 750 letters, with 90% opposing the proposed regulations as overly intrusive and punitive (Anderson, 1994). When the new rules became active in 1995, they only required disclosure of interests "that would reasonably appear to be directly and significantly affected by the research." Unfortunately, the public who would benefit from greater independence of science does not seem to have weighed in on this process, and the academic institutions receiving the grants ended up enforcing the regulations themselves (Witt & Gostin, 1994).

However, these initial steps seemed to have had little effect in controlling the industry's growing influence over medicine and the culture of universities (Witt & Gostin, 1994). In 1999, the American Society of Gene Therapy (ASGT) was forced to declare certain financial arrangements off-limits in gene therapy trials, following a scandal in the first gene therapy clinical trial (Therapy, 2000). Nonetheless, industry financing continued to dominate biomedicine, a trend which became clear in 1999 when the National Institutes of Health funded $17.8 billion for mostly basic research. In contrast, the leading 10 pharmaceutical companies spent $22.7 billion, on mostly clinical research (DeAngelis, 2000).

A spate of studies throughout the 1990s persisted in documenting corporate control over medicine. Research found that pharmaceutical companies affected clinicians' decisions (Chren & Landefeld, 1994) and that the research of academics with ties to industry was lower in quality (Bero & Rennie, 1996; Rochon, Gurwitz, Cheung, Hayes, & Chalmers, 1994) and more likely to favor the study sponsor's product (Cho & Bero, 1996; Rochon, Gurwitz, Simms, et al., 1994; Stelfox, Chua, O'Rourke, & Detsky, 1998). Negative findings were less likely to be published (Blumenthal, Campbell, Anderson, Causino, & Louis, 1997; Rennie, 1997) and more likely to have delayed publication (DeAngelis, 2000; Friedberg, Saffran, Stinson, Nelson, & Bennett, 1999). Especially worrying to academics was the media's growing interest in stories that documented industry's influence over medicine (Cohen, 2000; Peterson, 2000).

While the Bayh–Dole Act generated profits for universities and academics, it also constructed a positive feedback loop, driving more academic research down an avenue of commercialization. Whatever boundaries between universities and industry that had previously existed seemed to have disappeared as academic interests became almost indistinguishable from corporate interests (Korn, 2000). But the public's demand for advanced medical discoveries was tempered by intolerance for even a whiff of impropriety by universities now firmly entangled in corporate research. A JAMA editorial described this as a struggle "to create a precarious equipoise between the world and values of commerce and those of traditional public service, a balance between Bayh–Dole and by-God" (Korn, 2000).

Conflicts of interest captured attention again in 2000 when *USA Today* published an investigation that found that more than half of the advisors to the Food and Drug Administration (FDA) had financial relationships with pharmaceutical companies with interests in FDA decisions. Industry denied that these relationships created a problem and the FDA kept many of the financial details secret (Cauchon, 2000). A separate study found that companies funded almost one of every three manuscripts published in NEJM and JAMA. An expert concluded that financial conflicts of interest "is widespread among the authors of published manuscripts and these authors are more likely to present positive findings" (Friedman & Richter, 2004).

In retrospect, 2000 was a watershed event at JAMA. That year, the journal published a series of editorials examining the pharmaceutical industry's growing influence over physicians and called for barriers to protect medicine from corporate corruption. One editor noted that the industry's cultivations of physicians began in the first year of medical school when students received gifts from pharmaceutical companies (DeAngelis, 2000). "The enticement begins very early in a physician's career: for my classmates and me, it started with black bags," she wrote (DeAngelis, 2000). The editor referenced one study which found that pharmaceutical companies fund purportedly "independent physicians" and that research found that those academics were more likely to present positive findings (Friedman & Richter, 2004).

A steady trickle of research in the 2000s continued documenting widespread conflicts of interest which eroded scientific integrity, and explored disclosure as a primary tool for remediation (Bekelman, Li, & Gross, 2003; Friedman, 2002). However, one study discovered that barely half of the biomedical journals had policies requiring disclosure of conflicts of interest (Bekelman et al., 2003). Research also noted that companies appeared to be sponsoring studies as a tool to attack competitors' products and these studies were likely being funded for commercial not scientific reasons (L. S. Friedman & Richter, 2004). Management of conflicts of interest remained erratic, and a systematic review of journals, found that they were increasingly adopting disclosure policies, but those policies varied widely across disciplines, with medical journals more likely to have rules (Ancker & Flanagin, 2007). In response to that environment, the Natural Resources Defense Council convened a meeting and released a report on strengthening conflict of interest rules at journals (Sass, 2008).

Government investigations in the mid to late 2000s forced more biomedical conflicts of interest scandals onto the public stage. After the *Los Angeles Times* reported that some researchers at the National Institutes of Health had lucrative consulting agreements with industry, Congress held hearings, resulting in a tightening of conflict of interest policies for NIH employees (Willman, 2005). Federal investigations also began to force drug companies to disclose their payments to physicians on publicly available websites as part of corporate integrity agreements (*Report From February 23, 2012, Pharmaceutical Compliance Roundtable*, 2012).

Merck's Vioxx scandal threw a spotlight on the pharmaceutical industry's abuse of medical research in 2007. Documents made public during litigation found that Merck transformed peer-reviewed research into marketing brochures by ghostwriting studies for academics who rarely disclosed their industry ties (Psaty & Kronmal, 2008; Ross, Hill, Egilman, & Krumholz, 2008). Analyzing published articles, information Merck provided to the Food and Drug Administration, and Merck's internal analysis, researchers found that Merck may have misrepresented the risk–benefit profile of Vioxx in clinical trials and attempted to minimize mortality risk in reports to the FDA. For one trial, company documents revealed that the lack of a data and safety monitoring board (DSMB) may have endangered patients (Psaty & Kronmal, 2008). Lest anyone think that Merck was somehow unique in behavior, a JAMA editorial accompanying the papers referenced similar actions by other companies (Healy & Cattell, 2003; Silverstein et al., 2000; Solomon & Avorn, 2005; Steinman, Bero, Chren, & Landefeld, 2006). The editorial concluded, "[M]anipulation of study results, authors, editors, and reviewers is not the sole purview of one company" (DeAngelis & Fontanarosa, 2008).

In 2009, the Institute of Medicine (IOM) examined financial conflicts of interest in biomedicine, including research, education, and clinical practice (*Conflict of Interest in Medical Research, Education, and Practice*, 2009). The IOM reported that companies paid large, undisclosed amounts to doctors to give marketing talks to colleagues, and that sales representatives provided gifts to physicians that influence prescribing (Carlat, 2007; Elliott, 2006). Clinical research with unfavorable results was sometimes not published, distorting the scientific literature for drugs prescribed for arthritis, depression, and elevated cholesterol levels (Gibson, 2004; Kastelein et al., 2008; Whittington et al., 2004; Wright, Perry, Bassett, & Chambers, 2001). In one example, negative studies about depression medications were withheld (Healy, 2006; Turner, Matthews, Linardatos, Tell, & Rosenthal, 2008) causing a meta-analysis of the literature to find the drugs were safe and effective (*Preliminary Report of the Task Force on SSRIs and Suicidal Behavior in Youth*, 2004). A second meta-analysis that included the formerly withheld data found that risks outweighed benefits for all but one antidepressant (Whittington et al., 2004).

A fair reading of the IOM's report would cause any reader to conclude that conflicts of interest are pervasive throughout medicine, corrupt academia, and sometimes lead to patient harm. One expert has argued that polices to stop bias and corruption have been completely ineffective, requiring nothing less than a paradigm change in medicine's relationship with industry (Lexchin, 2012). Still, some research has found that the public remains largely unconcerned about these matters (Hampson et al., 2006; Weinfurt et al., 2006).

1.6 Perpetual Denial Machine

The defensive response by academics to the 1971 National Academy's first conflict of interest policy and the 1990 proposed regulations by the National Institutes of Health remains common to this day. Every attempt to control financial conflicts of interest and push for great transparency in science has been criticized by the scientific community, which seems perpetually satisfied with whatever ethics happen to be in place.

For example, the NIH's 1990 proposed guidelines were roundly denounced by the scientific community, resulting in gentler guidelines that allowed universities to self-regulate (Korn, 2000). Even with these weakened rules, a researcher later wrote, "At the present time, federal employees working in federal laboratories are constrained by numerous conflict of interest restrictions" (Witt & Gostin, 1994). Because of this perceived harshness, the NIH director eased ethics policies for NIH employees in 1995 to increase recruitment of top scientists, by allowing federal workers to consults with industry (Steinbrook, 2004).

Rolling back these rules led to inevitable scrutiny in the form of a 2003 investigation by the *Los Angeles Times* that uncovered senior NIH scientists consulting with pharmaceutical companies, with one researcher later prosecuted by the Department of Justice (Willman, 2003). Congressional hearings and internal investigations then forced the NIH to introduce more stringent ethics rules for employees that restricted stock ownership and consulting with pharmaceutical companies (Marshall, 2003). Announcing the new restrictions, the NIH Director stated a need to "preserve the public's trust" and address public perceptions regarding conflicts of interest (Kaiser, 2005). But as earlier, some scientists saw this second round of rules as punitive and overly restrictive, arguing that it would negate the agency's ability to recruit top scientists (Twombly, 2007).

Indeed, academics persisted in involving themselves in research that tested their own company's products on patients. In 2008, the Senate Finance Committee discovered that a Stanford University researcher had $6 million in equity in a company and was the primary investigator for an NIH grant that funded patient research on his company's drug (Krieger, 2008). Stanford denied any wrongdoing while also retaining a financial interest in the company. The NIH later terminated the clinical trial (Thacker, 2011).

Investigations by the Senate Finance Committee also uncovered numerous examples of academics failing to report financial ties to pharmaceutical companies when receiving NIH grants (Wadman, 2009). This led to reforms that required stronger conflicts of interest rules for NIH grantees and passage of the Physician Payments Sunshine Act (Thacker, 2013). The Sunshine Act, which I helped to write and pass, required companies to report payments to physicians, and the law has been replicated in many other countries. Despite the legislative success, the welcome in academia has been colder. In one example, Tufts University disinvited me from appearing at a conference on conflicts of interest held on their campus, which led one conference organizer to resign (Kowalczyk, 2009). Since these changes were implemented, industry and academia have attempted to rollback both provisions of the Sunshine Act and the new NIH rules (Reardon, 2015; Thacker, 2016a).

The Food and Drug Administration has had equally erratic responses to conflicts of interest. In 1999, a gene transfer experiment at the University of Pennsylvania killed volunteer patient Jesse Gelsinger. Both the investigator and the institution had financial interests in the tested product (Korn, 2000). The FDA then instituted more stringent conflicts of interest disclosure requirements for researchers and forbade those dealing with patients from holding equity, stock options or comparable arrangements in companies sponsoring the trial (Sibbald, 2001). Years after the incident, Gelsinger's father wrote, "So, my son, doing the right thing, was killed by a system and people rife with conflicts of interest, and real justice has been found to be very lax. It's essentially business as usual" (Administrator, 2008).

Driven in part by the Vioxx scandal, the FDA commissioned a 2006 study by the Institute of Medicine. That report found excessive conflicts of interest on the FDA's expert advisory panels which review new drugs and devices. The report recommended that a majority of the panelists should have no ties to the industry. "FDA's credibility is its most crucial asset, and recent concerns about the independence of advisory committee members… have cast a shadow on the trustworthiness of the scientific advice received by the agency," the report concluded (Vedantam, 2006).

In 2007, Congress responded, passing a new law that updated the Food, Drug, and Cosmetic Act that placed more stringent requirements on how the FDA handled conflicts of interest (Davis, 2014). In classic fashion, a senior FDA official later protested that the rules were harming the agency's ability to find qualified experts for advisory panels. These claims were rebutted in a letter to the FDA Commissioner, citing evidence that nearly 50% of research academics have no ties to industry and that approximately one-third of these researchers are full professors (Brian, 2011; Zinner, Bolcic-Jankovic, Clarridge, Blumenthal, & Campbell, 2009). Nonetheless the FDA outcry appeared effective and when Congress updated FDA legislation in 2012, the new law removed the previous demands that the FDA tighten control of financial conflicts of interest (Davis, 2014).

Even the journals themselves have joined the receding tide in handling conflicts of interest. After implementing the first conflict of interest policy in 1984, the NEJM updated its policies in 1990, prohibiting the authors of editorials and review articles from having any financial interests with a company that could benefit from a drug or medical device discussed in the article (Relman, 1990). The new rules created a firestorm of protest, with some calling them "McCarthyism" and others referring to them as "censorship" (Wilson, 2016). Eventually, the rules were weakened. Under a new editor in 2015, the NEJM published a series of editorials that sought to deny that conflicts of interest corrupt science (Brownlee, 2015).

Finally, another avenue for disclosing hidden conflicts of interest between industry and public scientists is through open records requests. Federal or state freedom of information laws enable investigative journalists and others to request documents relating to the publicly funded activity of many kinds, including scientific research (Seife, 2015). But in recent years, those laws have come under attack by the Union of Concerned Scientists and some members of the scientific community (Thacker, 2017c). Experts on freedom of information laws have dismissed these efforts as misguided, with one scholar referring to them as "gibberish" (Abel, 2016; Thacker, 2016c).

Even if compliance with current public records laws remains intact, the number of journalists using this tool is not large and is declining. In recent years, many journalists have also gone to work for the industries they once reported on (Thacker, 2016b). And like medicine, journalism has struggled with conflicts of interest problems, with most media outlets lacking clear policies for both reporters and the sources they cite (Thacker, 2015b). The Physicians Payments Sunshine Act has been used to uncover doctors, who are also reporters and who have received compensation from the pharmaceutical industry (Thacker, 2015a). And just as in science, the pharmaceutical, food, and biotech industries have secretly funded journalists to attend conferences on subjects that they cover in order to sway public perception (Thacker, 2017a, 2017b).

1.7 Discussion: Endless Search for Solutions

This brief history of financial conflicts of interest only attempts to examine the direct lineage that begins with tobacco, tracing it to modern problems in biomedicine. Other examples exist in which corporations sought to undermine scientific integrity for financial gain, but there is little evidence that those efforts continued into the future. History is important because it explains why these campaigns began, how they were implemented, and the tactics they deployed.

Historical wisdom also makes clear that reform efforts are always opposed, erode over time, and are then implemented again in the face of new scandals. As I was writing this chapter, the National Academies is implementing new conflicts of interest rules to deal with scandals involving two of their panels that had been stacked with academics who had ties to industry (Taylor, 2017). Additionally, the National Institutes of Health has been swept up in another controversy, with NIH officials soliciting donations from alcoholic beverage manufacturers to fund a $100 million study on the health effects of alcohol (Rabin, 2018). The NIH later ended the partnership (NIH to end funding for Moderate Alcohol and Cardiovascular Health trial,

2018). The resulting criticism seems to have curbed the NIH from partnering with the pharmaceutical industry on a planned opioids research partnership worth roughly $400 million, in which industry would fund half the costs (Facher, 2018).

The Institute of Medicine's 2009 report noted that the current evidence base for conflict of research policies is not strong and more research on the matter could help guide future rules or regulations (*Conflict of Interest in Medical Research, Education, and Practice*, 2009). Federal agencies have not leaped at this recommendation. The judicial branch may be more promising. Federal settlements with drug companies have forced them to disclose their payments to physicians and private litigation has uncovered documents showing bias in purportedly independent scientific studies (*Conflict of Interest in Medical Research, Education, and Practice*, 2009). The Senate has proposed the Sunshine in Litigation Act, which would require judges to make public documents that find products might harm the public, but this law has not been passed (Editors, 2014).

Tiny advances continue as PubMed announced it will include conflicts of interest statements with study abstracts, and research on the subject continues, even if the results are often ignored (Cronin, 2017). Searching PubMed for the term "conflict of interest" in 2006, a researcher found 4,623 entries with only 240 appearing before 1990, and well more than half after 1999 (Parascandola, 2007).

Most fixes for conflicts of interest involve some type of funding disclosure. But even these can be ineffective and distracting as disclosure does not resolve or eliminate the problem. Institutions must also evaluate and act on this information in ways that include eliminating the relationship or restricting a scientist's participation in some activities (*Conflict of Interest in Medical Research, Education, and Practice*, 2009).

Yet, some experts still attempt to dismiss the problem with conflicts of interest, by recasting the term as "confluence of interest" (Cappola & FitzGerald, 2015). Others trivialize the matter by elevating so-called "intellectual conflicts of interest" as similar in value (Parascandola, 2007). The Institute of Medicine carefully rejected such notions, stating, "Although other secondary interests may inappropriately influence professional decisions and additional safeguards are necessary to protect against bias from such interests, financial interests are more readily identified and regulated." The report concluded, "Such conflicts of interest threaten the integrity of scientific investigations, the objectivity of medical education, the quality of patient care, and the public's trust in medicine" (*Conflict of Interest in Medical Research, Education, and Practice*, 2009).

Many scientists are incapable of understanding and accepting that financial conflicts of interest corrupt science because they believe that scientists are objective and too well trained to be influenced by financial rewards, like all other human beings. In one example, researchers surveyed medical residents and found that sixty-one percent reported that they would *not* be influenced by gifts from pharmaceutical companies, while arguing that eighty-four percent of their colleagues *would* be influenced (Steinman, 2001). One academic who researches conflicts of interest grew so irritated with scientists denying the science of financial influence that he wrote a parody that listed many of their most common denials. "What I find most frustrating is the extent to which leading physicians and scientists whose profession seems to require a commitment to some kind of evidence based practice are unaware of the best evidence on motivated bias", he wrote. "This literature is robust and well developed" (Goldberg, 2015). Indeed, it is time for scientists to stop being unscientific about the science on conflicts of interest and to cease substituting their personal opinions for peer-reviewed research (Thacker, 2018).

A wide range of other industries have carefully studied the tobacco industry playbook. As a result, they have come to better understand the fundamentals of influence within the sciences and the value of uncertainty and skepticism in deflecting regulation, defending against litigation, and maintaining credibility despite marketing products that are known to harm public health (Brandt, 2012). "By making science fair game in the battle of public relations, the tobacco industry set a destructive precedent that would affect future debates on subjects ranging from global warming to food and pharmaceuticals" (Brownell & Warner, 2009).

At the heart of the matter lies money. As far back as 2000, experts questioned the ability of academic institutions to regulate financial conflicts of interests when they were so reliant on billions of dollars annually from the industry (Boyd & Bero, 2000). In a 2012 symposium on conflicts of interest held at Harvard Law School, academic leaders noted that the problem has only grown more and more complex over time (*Institutional Financial Conflicts of Interest in Research Universities, Symposium Report*, 2012). University leaders avoid even discussing the imperative to regulate financial conflicts because they fear losing revenue. Brave policymakers must intervene and develop rules to avoid future scandals and continued loss of confidence in science. Most importantly, they must protect the public.

References

Abel, D. (2016, March 19). How public must science be? Union of Concerned Scientists would limit disclosures. *Boston Globe*.

Academies, T. N. (2003). *Conflicts of interest policy for committees used in the development of reports*. Online.

Administrator. (2008). *A comment from Paul Gelsinger on gene therapy and informed consent*. Retrieved from http://www.bioethics.net/2008/01/a-comment-from-paul-gelsinger-on-gene-therapy-and/.

Ancker, J. S., & Flanagin, A. (2007). A comparison of conflict of interest policies at peer-reviewed journals in different scientific disciplines. *Science and Engineering Ethics, 13*(2), 147–157. https://doi.org/10.1007/s11948-007-9011-z.

Anderson, C. (1994). Agencies set rules on financial disclosure. *Science, 265*(5169), 179–180.

Bekelman, J. E., Li, Y., & Gross, C. P. (2003). Scope and impact of financial conflicts of interest in biomedical research: A systematic review. *JAMA, 289*(4), 454–465.

Bero, L. A., & Rennie, D. (1996). Influences on the quality of published drug studies. *International Journal of Technology Assessment in Health Care, 12*(2), 209–237.

Blumenthal, D., Campbell, E. G., Anderson, M. S., Causino, N., & Louis, K. S. (1997). Withholding research results in academic life science. Evidence from a national survey of faculty. *JAMA, 277*(15), 1224–1228.

Boffey, P. (1975). *The Brain Bank of America: An inquiry into the politics of science*. New York, NY: McGraw-Hill.

Boyd, E. A., & Bero, L. A. (2000). Assessing faculty financial relationships with industry: A case study. *JAMA, 284*(17), 2209–2214.

Brandt, A. M. (2007). *The cigarette century: The rise, fall, and deadly persistence of the product that defined America*. New York, NY: Basic Books.

Brandt, A. M. (2012). Inventing conflicts of interest: A history of tobacco industry tactics. *American Journal of Public Health, 102*(1), 63–71. https://doi.org/10.2105/AJPH.2011.300292.

Brian, D. (2011). [Letter to FDA Commissioner, Margaret Hamburg, via Email].

Brownell, K. D., & Warner, K. E. (2009). The perils of ignoring history: Big tobacco played dirty and millions died. How similar is big food? *Milbank Q, 87*(1), 259–294. https://doi.org/10.1111/j.1468-0009.2009.00555.x.

Brownlee, S. (2015). *The conflict denialists strike back*. Retrieved from http://lowninstitute.org/news/the-conflict-denialists-strike-back/.

Burling, C. (1963). *Public relations in the field of smoking and health*. Industry Documents. Retrieved April 30, 2018 https://www.industrydocumentslibrary.ucsf.edu/tobacco/docs/htfl0012.

Cappola, A. R., & FitzGerald, G. A. (2015). Confluence, not conflict of interest: Name change necessary. *JAMA, 314*(17), 1791–1792. https://doi.org/10.1001/jama.2015.12020.

Carlat, D. (2007, November 25). Dr. Drug Rep. *New York Times*.

Cauchon, D. (2000, September 25). FDA advisers tied to industry. *USA Today*.

Centers, A. o. A. H. (1990). *Conflicts of interest in academic health centers*. Washington, DC.

Cho, M. K., & Bero, L. A. (1996). The quality of drug studies published in symposium proceedings. *Annals of Internal Medicine, 124*(5), 485–489.

Chren, M. M., & Landefeld, C. S. (1994). Physicians' behavior and their interactions with drug companies. A controlled study of physicians who requested additions to a hospital drug formulary. *JAMA, 271*(9), 684–689.

Cohen, R. (2000, September 24, 2000). Conference call. *New York Times Magazine*.

Cohn, V. (1971, April 26). Science academy facing cries for reform. *Washington Post*.

Colleges, A. o. A. M. (1990). *Guidelines for dealing with faculty conflicts of commitment and conflicts of interest in research*. Ethics Codes Collection.

Committee, T. I. R. (1954). *A frank statement to cigarette smokers*. Industry Documents. Retrieved April 30, 2018, from https://www.industrydocumentslibrary.ucsf.edu/tobacco/docs/zkph0129.

Conflict of Interest in Medical Research, Education, and Practice. (2009). Retrieved from.

Cronin, J. (2017). *PubMed to include conflict-of-interest statements with study abstracts* [Press release]. Retrieved from https://cspinet.org/news/pubmed-include-conflict-interest-statements-study-abstracts-20170418.

Culliton, B. J. (1982). The academic-industrial complex. *Science, 216*(4549), 960–962.

Davis, C. (2014). Red tape tightrope: Regulating financial conflicts of interest in FDA advisory committees. *Washington University Law Review, 91*(6).

DeAngelis, C. D. (2000). Conflict of interest and the public trust. *JAMA, 284*(17), 2237–2238.

DeAngelis, C. D., & Fontanarosa, P. B. (2008). Impugning the integrity of medical science: The adverse effects of industry influence. *JAMA, 299*(15), 1833–1835. https://doi.org/10.1001/jama.299.15.1833.

Editors. (2014, May 31). Secrecy that kills. *New York Times*.

Elliott, C. (2006). The drug pushers. *Atlantic Magazine*.

Facher, L. (2018, April 19). NIH abruptly changes course on industry opioids partnership after ethics flags raised. *STAT*. Retrieved from https://www.statnews.com/2018/04/19/nih-industry-partnership-ethics-opioids/.

Fauber, J. (2014, June 25). 9 of 19 experts on pain panel tied to drug companies. *Milwaukee Journal Sentinel*.

Friedberg, M., Saffran, B., Stinson, T. J., Nelson, W., & Bennett, C. L. (1999). Evaluation of conflict of interest in economic analyses of new drugs used in oncology. *JAMA, 282*(15), 1453–1457.

Friedman, L. S., & Richter, E. D. (2004). Relationship between conflicts of interest and research results. *Journal of General Internal Medicine, 19*(1), 51–56.
Friedman, P. J. (2002). The impact of conflict of interest on trust in science. *Science and Engineering Ethics, 8*(3), 413–420.
Gibson, L. (2004). GlaxoSmithKline to publish clinical trials after US lawsuit. *BMJ, 328*(7455), 1513. https://doi.org/10.1136/bmj.328.7455.1513-a.
Gillette, R. (1972). Academy food committees: New criticism of industry ties. *Science, 177*(4055), 1172–1175. https://doi.org/10.1126/science.177.4055.1172.
Gingras, Y., & Gosselin, P. M. (2008). The emergence and evolution of the expression "conflict of interests" in science: A historical overview, 1880–2006. *Science and Engineering Ethics, 14*(3), 337–343. https://doi.org/10.1007/s11948-008-9063-8.
Goldberg, D. S. (2015). COI bingo BMJ opinion. https://blogs.bmj.com/bmj/2015/08/24/daniel-s-goldberg-coi-bingo/.
Guttman, D. W. B. (1976). *The shadow government: The government's multi-billion-dollar give-away of its decision-making powers to private management consultants, "experts," and think tanks*. New York, NY: Pantheon.
Hampson, L. A., Agrawal, M., Joffe, S., Gross, C. P., Verter, J., & Emanuel, E. J. (2006). Patients' views on financial conflicts of interest in cancer research trials. *New England Journal of Medicine, 355*(22), 2330–2337. https://doi.org/10.1056/NEJMsa064160.
Hanmer, H. (1954). *Re: Rewrite of "White Paper" introduction*. Industry Documents. Retrieved April 30, 2018, from Ness Motley Law Firm Documents. https://www.industrydocumentslibrary.ucsf.edu/tobacco/docs/ltyc0040.
Healy, D. (2006). Did regulators fail over selective serotonin reuptake inhibitors? *BMJ, 333*(7558), 92–95. https://doi.org/10.1136/bmj.333.7558.92.
Healy, D., & Cattell, D. (2003). Interface between authorship, industry and science in the domain of therapeutics. *British Journal of Psychiatry, 183*, 22–27.
Hill, J. (1962). *Smoking, health and statistics: The story of the tobacco accounts*. Industry Documents. Retrieved April 30, 2018, from Hill & Knowlton Inc. https://www.industrydocumentslibrary.ucsf.edu/tobacco/docs/tyfl0060.
Inc, H. K. (1953). *Preliminary recommendations for cigarette manufacturers*. Industry Documents. Retrieved April 30, 2018, from Unknown. https://www.industrydocumentslibrary.ucsf.edu/tobacco/docs/rsfy0137.
Inc, H. K. (1954). *Press release, re: Timothy V. Hartnett named as full-time chairman of Tobacco Industry Research Committee*. Retrieved from Hill & Knowlton Inc. https://www.industrydocumentslibrary.ucsf.edu/tobacco/docs/jxwb0035.
Inc, H. K. (1958). *Ex-Congressman Richards to Head Newly-Formed Tobacco Institute, Inc.*
Institutional Financial Conflicts of Interest in Research Universities, Symposium Report. (2012). Retrieved from Cambridge, MA. http://petrieflom.law.harvard.edu/assets/publications/fCOI_final_report.pdf.
Jasanoff, S. (1998). *The fifth branch: Science advisors as policy makers*. Boston, MA: Harvard University Press.
Jukes, T. H. (1983). Philip Handler (1917–1981). A biographical sketch. *The Journal of Nutrition, 113*(6), 1086–1094.
Kaiser, J. (2005). Conflict of interest. NIH chief clamps down on consulting and stock ownership. *Science, 307*(5711), 824–825. https://doi.org/10.1126/science.307.5711.824.
Kastelein, J. J., Akdim, F., Stroes, E. S., Zwinderman, A. H., Bots, M. L., Stalenhoef, A. F., … Investigators, E. (2008). Simvastatin with or without ezetimibe in familial hypercholesterolemia. *New England Journal of Medicine, 358*(14), 1431–1443. https://doi.org/10.1056/nejmoa0800742.
Klopsteg, P. E. (1955). Role of government in basic research. *Science, 121*(3153), 781–784. https://doi.org/10.1126/science.121.3153.781.
Korn, D. (2000). Conflicts of interest in biomedical research. *JAMA, 284*(17), 2234–2237.
Kowalczyk, L. (2009, March 26). Ethics event spawns a tussle at Tufts: Speaker is out; now organizer is, too. *Boston Globe*.

Krieger, L. (2008, June 24). Stanford doctor's stock raises ethics concerns. *The Mercury News.*
Lelyveld, J. (1963, November 29). Cigarette makers prosper despite debate on hazards. *New York Times.*
Lexchin, J. (2012). Those who have the gold make the evidence: How the pharmaceutical industry biases the outcomes of clinical trials of medications. *Science and Engineering Ethics, 18*(2), 247–261. https://doi.org/10.1007/s11948-011-9265-3.
Marshall, E. (2003). Conflict of interest. Zerhouni pledges review of NIH consulting in wake of allegations. *Science, 302*(5653), 2046. https://doi.org/10.1126/science.302.5653.2046.
Michael, D. (2008, May 1). *Doubt is their product: How industry's assault on science threatens your health* (1st ed.). Oxford University Press. https://www.amazon.es/Doubt-Their-Product-Industrys-Threatens/dp/019530067X.
Nelson, B. (1968). Philip Handler: National Academy nominates a worldly "high priest". *Science, 162*(3857), 981–984.
NIH to end funding for Moderate Alcohol and Cardiovascular Health trial (2018). NIH to end funding for Moderate Alcohol and Cardiovascular Health trial. https://www.nih.gov/news-events/news-releases/nih-end-funding-moderate-alcohol-cardiovascular-health-trial.
Oreskes, N. C. E. (2010). *Merchants of doubt: How a handful of scientists obscured the truth on issues from tobacco smoke to global warming*. Bloomsbury Press.
Parascandola, M. (2004). Skepticism, statistical methods, and the cigarette: A historical analysis of a methodological debate. *Perspectives in Biology and Medicine, 47*(2), 244–261.
Parascandola, M. (2007). A turning point for conflicts of interest: The controversy over the National Academy of Sciences' first conflicts of interest disclosure policy. *Journal of Clinical Oncology, 25*(24), 3774–3779. https://doi.org/10.1200/JCO.2006.09.2890.
Parascandola, M., Weed, D. L., & Dasgupta, A. (2006). Two Surgeon General's reports on smoking and cancer: A historical investigation of the practice of causal inference. *Emerging Themes in Epidemiology, 3,* 1. https://doi.org/10.1186/1742-7622-3-1.
Peterson, M. (2000, August 27). What's black and white and sells medicine? *New York Times.*
Preliminary Report of the Task Force on SSRIs and Suicidal Behavior in Youth. (2004). Retrieved from Executive Summary. https://acnp.org/wp-content/uploads/2017/10/Task-Force-Report-on-SSRIs-and-Suicide-in-Youth-Exec-Summary.pdf.
Press, E. W. J. (2000). The kept university. *Atlantic Monthly, 284,* 2234–2237.
Proctor, R. S. L. (2008). *Agnotology: The making and unmaking of ignorance*. Stanford University Press.
Professors, A. A. o. U. (1915). Declaration of principles. *Bulletin of the American Association of University Professors, 1*(1).
Psaty, B. M., & Kronmal, R. A. (2008). Reporting mortality findings in trials of rofecoxib for Alzheimer disease or cognitive impairment: A case study based on documents from rofecoxib litigation. *JAMA, 299*(15), 1813–1817. https://doi.org/10.1001/jama.299.15.1813.
Rabin, R. (2018, March 17). Federal agency courted alcohol industry to fund study on benefits of moderate drinking. *New York Times.*
Reardon, S. (2015). NIH disclosure rules falter. *Nature, 525*(7569), 300–301. https://doi.org/10.1038/525300a.
Relman, A. S. (1984). Dealing with conflicts of interest. *New England Journal of Medicine, 310*(18), 1182–1183. https://doi.org/10.1056/NEJM198405033101809.
Relman, A. S. (1990). New "Information for Authors"—And readers. *New England Journal of Medicine, 323*(1), 56. https://doi.org/10.1056/NEJM199007053230111.
Rennie, D. (1997). Thyroid storm. *JAMA, 277*(15), 1238–1243.
Report From February 23, 2012, Pharmaceutical Compliance Roundtable. (2012). Retrieved from https://oig.hhs.gov/compliance/compliance-guidance/docs/pharmaceutical-compliance-roundtable.pdf.
Rochon, P. A., Gurwitz, J. H., Cheung, C. M., Hayes, J. A., & Chalmers, T. C. (1994). Evaluating the quality of articles published in journal supplements compared with the quality of those published in the parent journal. *JAMA, 272*(2), 108–113.

Rochon, P. A., Gurwitz, J. H., Simms, R. W., Fortin, P. R., Felson, D. T., Minaker, K. L., & Chalmers, T. C. (1994). A study of manufacturer-supported trials of nonsteroidal anti-inflammatory drugs in the treatment of arthritis. *Archives of Internal Medicine, 154*(2), 157–163.

Ross, J. S., Hill, K. P., Egilman, D. S., & Krumholz, H. M. (2008). Guest authorship and ghostwriting in publications related to rofecoxib: A case study of industry documents from rofecoxib litigation. *JAMA, 299*(15), 1800–1812. https://doi.org/10.1001/jama.299.15.1800.

Sass, J. (2008). *Effective and practical disclosure policies: NRDC paper on workshop to identify key elements of disclosure policies for health science journals*. Retrieved from Washington, DC. https://www.nrdc.org/sites/default/files/disclosure.pdf.

School, H. M. (2010). *Changes to conflicts policy heighten transparency*. Online.

Scientists in Government: Growing Concern Over Conflicts of Interest. (1960). *Science, 131*(3412), 1508–1509. https://doi.org/10.1126/science.131.3412.1508.

Seife, C. T. P. (2015, August 21). Why it's OK for taxpayers to 'snoop' on scientists. *Los Angeles Times*.

Sibbald, B. (2001). Death but one unintended consequence of gene-therapy trial. *CMAJ, 164*(11), 1612.

Silverstein, F. E., Faich, G., Goldstein, J. L., Simon, L. S., Pincus, T., Whelton, A., … Geis, G. S. (2000). Gastrointestinal toxicity with celecoxib vs nonsteroidal anti-inflammatory drugs for osteoarthritis and rheumatoid arthritis: The CLASS study: A randomized controlled trial. Celecoxib long-term arthritis safety study. *JAMA, 284*(10), 1247–1255.

Solomon, D. H., & Avorn, J. (2005). Coxibs, science, and the public trust. *Archives of Internal Medicine, 165*(2), 158–160. https://doi.org/10.1001/archinte.165.2.158.

Steinbrook, R. (2004). Financial conflicts of interest and the NIH. *New England Journal of Medicine, 350*(4), 327–330. https://doi.org/10.1056/NEJMp038247.

Steinman, M. A. (2001). Of principles and pens: attitudes and practices of medicine housestaff toward pharmaceutical industry promotions. In M. A. Steinman, M. G. Shlipak & S. J. McPhee (Eds.), *The American Journal of Medicine, 110*(7), 551–557. https://www.ncbi.nlm.nih.gov/pubmed/11347622.

Steinman, M. A., Bero, L. A., Chren, M. M., & Landefeld, C. S. (2006). Narrative review: The promotion of gabapentin: An analysis of internal industry documents. *Annals of Internal Medicine, 145*(4), 284–293.

Stelfox, H. T., Chua, G., O'Rourke, K., & Detsky, A. S. (1998). Conflict of interest in the debate over calcium-channel antagonists. *New England Journal of Medicine, 338*(2), 101–106. https://doi.org/10.1056/NEJM199801083380206.

Strom, S. (2016, December 27). National biotechnology panel faces new conflict of interest questions. *New York Times*.

Taylor, A. (2017, May 3). National academies revise conflict of interest policy. *The Scientist*.

Thacker, P. (2011). *The ugly underbelly of medical research*. Retrieved from http://pogoblog.typepad.com/pogo/2011/01/the-ugly-underbelly-of-medical-research.html.

Thacker, P. (2013). *The slow pace of success in a 'do something congress'*. Retrieved from https://ethics.harvard.edu/blog/slow-pace-success-do-something-congress.

Thacker, P. (2014, December 2). Consumers deserve to know who's funding health research. *Harvard Business Review*.

Thacker, P. (2015a). Database may uncover conflicts of interest for TV doctors. *Columbia Journalism Review*.

Thacker, P. (2015b, November 23). Where do science journalists draw the line? *Columbia Journalism Review*.

Thacker, P. (2016a, August 10). Don't rewrite history: 'Sunshine' law was meant to protect docs from undue influence. *STAT*. Retrieved from https://www.statnews.com/2016/08/10/sunshine-act-revise-history/.

Thacker, P. (2016b). How a former reporter is helping big oil and gas frack the news: In the heated fight over fracking, industry gets insider tips on shaping its story. *HuffPo*. Retrieved from https://www.huffingtonpost.com/entry/colorado-fracking-free-press_us_57f565eae4b05f39c51e0916.

Thacker, P. (2016c, January 9). Scientists, give up your emails. *New York Times*.
Thacker, P. (2017a). Coca-Cola's secret influence on medical and science journalists. *BMJ, 357*(j1638). https://doi.org/10.1136/bmj.j1638.
Thacker, P. (2017b, July 21). Flacking for GMOs: How the biotech industry cultivates positive media—And discourages criticism. *The Progressive*.
Thacker, P. (2017c, March 21). Why scientific transparency is so tricky. *Pacific Standard*.
Thacker, P. (2018, October 5). Time to act on industry influence over medicine and lay zombie arguments to rest. We have the evidence that change is needed, says Paul D Thacker. https://blogs.bmj.com/bmj/2018/10/05/paul-d-thacker-time-to-act-on-industry-influence-over-medicine-and-lay-zombie-arguments-to-rest/.
Therapy, A. S. o. G. (2000). *Policy of the American Society of Gene Therapy on financial conflict of interest in clinical research*. Milwaukee, WI.
Thompson, D. F. (1993). Understanding financial conflicts of interest. *New England Journal of Medicine, 329*(8), 573–576. https://doi.org/10.1056/NEJM199308193290812.
Thursby, J. G., & Thursby, M. C. (2003). Intellectual property. University licensing and the Bayh-Dole Act. *Science, 301*(5636), 1052. https://doi.org/10.1126/science.1087473.
Turner, E. H., Matthews, A. M., Linardatos, E., Tell, R. A., & Rosenthal, R. (2008). Selective publication of antidepressant trials and its influence on apparent efficacy. *New England Journal of Medicine, 358*(3), 252–260. https://doi.org/10.1056/NEJMsa065779.
Twombly, R. (2007). Conflict-of-interest rules worry some scientists. *Journal of the National Cancer Institute, 99*(1), 6–9. https://doi.org/10.1093/jnci/djk037.
Tye, L. (1998). *The father of spin: Edward L. Bernays & the birth of public relations* (1st ed.). New York: Crown Publishers.
United States v Philip Morris et al., No. Civil Action No. 99–2496 (GK) (2004).
Vedantam, S. (2006, September 23). FDA told U.S. drug system is broken. *Washington Post*.
Wadman, M. (2009). Money in biomedicine: The senator's sleuth. *Nature, 461*(7262), 330–334. https://doi.org/10.1038/461330a.
Weatherall, D. (2000). Academia and industry: Increasingly uneasy bedfellows. *Lancet, 355*(9215), 1574. https://doi.org/10.1016/S0140-6736(00)02211-X.
Weinfurt, K. P., Friedman, J. Y., Allsbrook, J. S., Dinan, M. A., Hall, M. A., & Sugarman, J. (2006). Views of potential research participants on financial conflicts of interest: Barriers and opportunities for effective disclosure. *Journal of General Internal Medicine, 21*(9), 901–906. https://doi.org/10.1111/j.1525-1497.2006.00502.x.
Whittington, C. J., Kendall, T., Fonagy, P., Cottrell, D., Cotgrove, A., & Boddington, E. (2004). Selective serotonin reuptake inhibitors in childhood depression: Systematic review of published versus unpublished data. *Lancet, 363*(9418), 1341–1345. https://doi.org/10.1016/S0140-6736(04)16043-1.
Willman, D. (2003, December 7). Stealth merger: Drug companies and government research. *Los Angeles Times*. Retrieved from http://articles.latimes.com/2003/dec/07/nation/na-nih7.
Willman, D. (2005, February 2, 2005). NIH seeks 'higher standard'. *Los Angeles Times*.
Wilson, M. (2016). The New England Journal of Medicine: Commercial conflict of interest and revisiting the Vioxx scandal. *Indian Journal of Medical Ethics, 1*(3), 167–171. https://doi.org/10.20529/IJME.2016.046.
Witt, M. D., & Gostin, L. O. (1994). Conflict of interest dilemmas in biomedical research. *JAMA, 271*(7), 547–551.
Wright, J. M., Perry, T. L., Bassett, K. L., & Chambers, G. K. (2001). Reporting of 6-month vs 12-month data in a clinical trial of celecoxib. *JAMA, 286*(19), 2398–2400.
Zinner, D. E., Bolcic-Jankovic, D., Clarridge, B., Blumenthal, D., & Campbell, E. G. (2009). Participation of academic scientists in relationships with industry. *Health Affairs (Millwood), 28*(6), 1814–1825. https://doi.org/10.1377/hlthaff.28.6.1814.

Paul D. Thacker is a journalist who writes on science and medicine. He has reported on scientific ethics for numerous outlets including the New York Times, Los Angeles Times, Washington Post, JAMA, Salon, BMJ, NEJM, The New Republic, and Mother Jones. He is aformer Fellow at the Edmond J. Safra Center for Ethics at Harvard and has given talks at numerous universities on journalism, integrity in science and biomedicine. For several years, he worked a congressional staffer and led a series of high profile investigations for the United States Senate that led to reforms in medicine, including passage of the Physician Payments Sunshine Act and reforms to the conflicts of interest policy of the National Institutes of Health.

Chapter 2
The Health Impact Fund and the Problem of Corruption in the Global Pharmaceutical Sector

Fred Gifford

Abstract The goal of this chapter is to examine the Health Impact Fund, a proposal aimed at solving the problem of access to pharmaceuticals for the global poor, focusing on its potential implications vis-à-vis the problem of corruption in the pharmaceutical industry. One set of questions concerns the extent to which the Health Impact Fund can be expected to help lessen or prevent the sorts of corruption that presently plague our system of pharmaceutical development. The second set of questions concerns the extent to which the sorts of structures and processes that are set up for the Health Impact Fund—making assessments of the impact of a given medicine on global disease burden—would themselves be vulnerable to corruption.

2.1 Introduction

Corruption concerning the development, evaluation, and distribution of pharmaceuticals, and in the health sector more broadly, is a major problem. This corruption, and the system of laws, regulations, and opportunities for profit that shapes and sustains it brings serious harm, including high prices, lack of availability of certain important medicines, and a distorted medical knowledge-base. The profit motive and power of the drug companies make reining in this corruption, and the harm and injustice that results, very difficult. Progress here will plausibly require a multipronged approach. The goal of this chapter is to examine the Health Impact Fund, a proposal aimed at solving the problem of access to pharmaceuticals for the global poor (rather than on corruption per se), focusing on its implications vis-à-vis the problem of corruption.

F. Gifford (✉)
Michigan University, Ann Arbor, USA
e-mail: gifford@msu.edu

K. T. Çalıyurt (ed.), *Integrity, Transparency and Corruption in Healthcare & Research on Health, Volume I*, Accounting, Finance, Sustainability, Governance & Fraud: Theory and Application, https://doi.org/10.1007/978-981-15-1424-1_2

2.1.1 The Health Impact Fund

The Health Impact Fund is an innovative proposal for counteracting the perverse incentives inherent in what Pogge and Hollis call the present patent regime, and for incentivizing instead the development of medicines aimed at diseases prevalent in the developing world (Hollis & Pogge, 2008). It would work by creating an additional, alternative track for drug companies to seek profit from the creation of a drug. For any given new drug, a pharmaceutical company could choose between registering it with the traditional patent system, or instead registering it with this new (HIF) system. If this alternate track is chosen for a particular drug, then that drug is to be made available, everywhere, at the lowest feasible cost of production and distribution. In return, the company is rewarded by receiving a portion of the HIF funds, in proportion to the impact demonstrated by this drug on the global disease burden over a period of, say, 10 years. Companies are thus rewarded on the basis of their products' impact on health, rather than on the basis of sales.

The proposal is quite controversial; one may worry that it is not feasible. But it should be explored because it is in many ways attractive. As well as leading to the availability of drugs for the global poor, it could be expected to help to ameliorate several other problems of the present patent-oriented drug development regime.

The focus of this essay is the implications of this HIF proposal for corruption in the pharmaceutical industry. In what follows, I first note the set of standard sorts of corrupt activities of drug companies. I then outline more broadly the failures of the drug development system in terms of injustice and inefficiencies, and note how the HIF can address these. Finally, I move on to this essay's main questions, concerning how the Health Impact Fund may be expected to impact issues of corruption.

2.1.2 Corrupt Activities of Pharmaceutical Companies

Pharmaceutical companies engage in a variety of corrupt activities in their quests to get their products through formal governmental drug approval processes, to convince physicians to prescribe their drugs, and to keep other companies from obtaining market share. These activities include the following:

1. the manipulation of trial results: calculated decisions about trial design concerning such things as powering of trials, eligibility criteria, and comparison treatment and dosages
2. failures to disclose such things as negative trial results and data about adverse reactions
3. manipulative promotion designed to change the beliefs and prescription practices of clinical practitioners: ghostwriting, the use of drug reps, free gifts, and lunches, and influence over continuing medical education
4. the buying of influence and loyalty, such as payments and perks to scientists and physicians

5. various unwarranted legal actions: "evergreening", legal intimidation of generic companies via threats to sue without real cause

Another kind of corruption concerns pharmaceutical companies' contributions to regulators and lawmakers, thereby influencing their regulatory and economic environment, and thus their ability to carry out each of the above.

These actions count as corruption because the agents take advantage of their position of power, including their ability generated by the lack of transparency, to gain illegitimate benefits for themselves (Crombie, 2020). (These actions are not necessarily illegal, but they do break a trust.)

2.1.3 Problems of Injustice and Inefficiency with the Present System

The actions of drug companies, and the consequences of those actions, take place within and are shaped by a set of features that make up the present patent regime. These include a set of regulatory and patent laws and economic and political forces, along with standards and expectations within the corporate and regulatory culture, which shape the motivations and abilities of the pharmaceutical companies to seek profits.

This system results in both injustices and inefficiencies (including further temptation or risk of corruption): First, drugs are too expensive. The poor can't afford the drugs that exist, because the prices are set at a level based on what the affluent can afford. But indeed, drug prices are very high for everyone, putting pressure on health-care systems in both rich and poor countries. (This is one reason why governments in the Global North and other entities could be expected to have more than just humanitarian (or global stability) reasons for being willing to contribute large amounts of money to the Health Impact Fund, hence making the system economically feasible.)

Second, there is an inappropriate bias in what drugs are developed in the first place. Drugs needed specifically in poor countries do not tend to be developed. Further, even for rich countries, there is too much bias toward "maintenance" drugs and "me-too" drugs, ones that contribute comparatively little, if at all, to improved health.

Third, the system encourages a particular form of corruption, the counterfeiting of drugs, threatening patient health and contributing to drug-specific resistance.

Fourth is the "last mile problem", that hoped-for benefits of a drug fail to materialize due to weak infrastructure (transport, storage, or an adequate health care system).

Finally, in many ways, this system of motivations and expectations leads to inefficiencies from the point of view of the drug companies themselves: There is a great deal of wasted effort on such practices as the multiple filing for patents, surveillance for infringement, and marketing. There are also "deadweight losses": many intermediate income countries would be willing to buy a drug at a lower but substantial price, but the system requires that it be sold only at the price set by the affluent, so these potential profits are lost.

The Health Impact Fund proposal addresses each of these problems. For instance, for drugs on the Health Impact Fund track, there would be no need for such a huge amount of resources to go toward litigation and surveillance for patent violations. And with prices so low, there would be no incentive to counterfeit. And importantly, concerning the "last mile" problem, the focus on health outcomes rather than sales would motivate innovators to take actions to ensure their medicine has a maximum health impact, for example, through the promotion of infrastructure to ensure medicines get distributed, and of adequate medical personnel to promote their proper use. Companies that contribute to these features would find their drugs making a larger impact, and would reap rewards from this.

One lesson here is that the Health Impact Fund schema has systemic effects. Amongst the impacts that ought to be examined are those it might have on addressing corruption.

2.1.4 Health Impact Fund as a "Re-incentivization" Strategy

The Health Impact Fund proposal is an instance of a broader category of strategy, namely the "re-incentivization" approach, involving the attempt to alter the reward structure, leading companies to take socially beneficial actions in the course of following the profit motive. (Note that another kind of proposal in this category would involve labeling products of a company based on their overall index of their social responsibility.) (Eyal, 2012).

This is a different strategy from "management" solutions, such as changes in standards or requirements of medical journals and regulatory agencies such as the FDA, with improved surveillance, policing, and stronger "conflict of interest" rules, as well as better enforcement of the rules we have now. It also contrasts with what might be called the "divestment" approach, (Brody, 2007) such as publicly funding pharmaceutical research rather than relying on pharmaceutical companies, on the grounds that there simply is no way to rein in their corruptive practices.

Of course, the idea is not that we have to choose the one right strategy here. These multiple strategies for dealing with corruption and bad outcomes concerning pharmaceutical development are, for the most part, compatible. Indeed, presumably no one strategy will be adequate on its own, so we need to take seriously a multipronged approach. Hence, when evaluating the Health Impact Fund's impact on corruption, this must be taken into account.

2.2 Overview of the Basic Questions of the Paper

So the question arises of how the Health Impact Fund would fare with respect to questions of corruption, or how corruption might be altered or affected by the HIF. In what follows, I consider two sets of questions about the Health Impact Fund proposal and its impact on corruption.

First, can the Health Impact Fund be expected to help lessen or prevent the sorts of corruption that presently plague our system of pharmaceutical development? (If so, then perhaps this counts in favor of the proposal, or at least gives us reason to study the proposal to see what can be learned from it.)

Second, might the sorts of processes involved in the Health Impact Fund themselves be vulnerable to corruption, thus raising *new* problems? (If so, perhaps this would cancel out the above benefits, or even make the Health Impact Fund not work after all.)

The Health Impact Fund is of course quite controversial, but it is worth exploring carefully both because its positive features make it worth pursuing as part of a solution, and also because the "re-incentivization" approach more broadly deserves exploration, even if we don't put in place the Health Impact Fund in particular.

Note also that the specific aim here is to explore the Health Impact Fund in relation to corruption, and not to provide a general defense of the Health Impact Fund. Nor is the aim here to provide empirical evidence about what would happen under a Health Impact Fund approach, but to lay out the issues in a way that presents a plausibility argument and directs us concerning what empirical evidence ought to be sought.

2.2.1 *Question 1: How Might the Health Impact Fund Reduce or Combat Corruption?*

Now, the arguments for thinking that the Health Impact Fund would help to diminish corruption are not that it is a mechanism for rooting out corruption (via the creation of new rules or their enforcement with better surveillance). Rather, the idea is that, for those drugs registered with the Health Impact Fund, it can remove or diminish certain motivations for these manipulative and deceptive practices. Within the Health Impact Fund framework, increased sales per se will not be an effective way to maximize profits, so the above endeavors will not be worth the effort.

For instance, within the Health Impact Fund track, prices are already determined at a low rate, so it would make no sense to use corrupt methods to try to keep the prices high. Nor would there be so much to gain from attempts either to bias the results of RCTs submitted to a drug approval agency, or to distort knowledge of clinicians and affect their prescribing habits. Rather, profits will be most effectively generated by making a product which genuinely (and measurably) addresses health problems, as well as by putting efforts and resources into addressing "last mile" solutions, instead of putting these efforts and resources into these corruptive practices.

Some caveats:
One caveat here is that there will only be an impact in this particular, small segment of the market, whereas the majority of drugs will continue to take the traditional patent track. So to evaluate this, we would need to know how widespread uptake of this practice will be. But even if the impact is small, it can be positive; further, it might serve as a demonstration of a process with less corruption that we can study and from which we can learn. Of course, its advocates might hope that over time this could become a larger and larger segment of the market. Ultimately, the empirical question of just how much of an impact this should be expected to have would be worth exploring.

Another caveat is this. Even within the Health Impact Fund path, it would not be accurate to say that companies simply cannot benefit at all from deception or other corruption. If we focus on those cases where the drug in fact works, companies can still be better off if there is more use of these products. More use will create more impact to be measured, thus increasing the credit the company will receive. Thus certain manipulative and deceptive methods for increasing use could still be in their interest, for increasing the number of instances of its use will increase its "overall impact". Hence, the companies could still be motivated to engage in various corrupt practices, such as ghostwriting, or designing a trial so that the drug looks better than it is.

But even if this is true, the Health Impact Fund schema will still have succeeded in significantly diminishing the advantages to be gained from such corruption, and will hence diminish the motivations for such "corruptive" activities, and hence lessen the incidence of corrupt actions themselves. For such activities will not be nearly as effective (and nearly as distorting) as in the standard regime. For there is only a motive to increase usage if the drug really will have a positive impact on the global disease burden (because effective and safe). This narrows the set of cases where it is done (unless the companies deceive themselves into believing their drug really is valuable when it is not). Further, the cases where it can still occur do not cause as much harm: it will not promote the use of ineffective or unsafe drugs, not will it damage our knowledge-base by misleading concerning how effective per dose the drug is. So the Health Impact Fund will have at least curbed some of the most problematic parts of the problem.

It is worth noting that in the important case of drugs for use in developing nations, it is likely to be the Ministries of Health (rather than individual doctors in the Global North) that one is trying to convince. Hence, the pharmaceutical companies' usual methods of sending drug reps to dispensing clinicians, etc., will not work. Of course, they may well find ways to manipulate the views of Ministries of Health. This then points us to a number of empirical questions that will require further examination, but which I must leave aside here: e.g., exactly how decisions would be made by Ministries of Health, and then, in turn, what strategies the drug companies could discover about how to influence this. In any case, the main point is that there will be much less motivation (or opportunity) to carry out the sorts of actions discussed above, and what ones remain will be less damaging.

A further and independent way in which the Health Impact Fund could be expected to diminish corruption concerns a particular potential benefit connected with the "last mile" problem. One of the central features of the last mile problem is this: Failure of a health intervention (like a drug) to achieve the health benefit hoped for may often be the result of poor or inefficient healthcare systems. The drug exists, and would be safe and effective if available and taken properly, but one or more obstacles to its success occur: It is not properly distributed to all parts of the health care system where it is needed. Or there aren't trained professionals to make judgments about and administer it. Or there are fees for the health care that keep many people away. Or the drug would be more effective if conjoined with better nutrition and avoidance of other diseases. Indeed, one important source of inefficiency in this system may be corruption in the healthcare system.

Thus an especially effective way to address the last mile problem will be to improve a healthcare system. But the presence of poor healthcare systems is, in fact, an important factor leading to increased corruption in that healthcare system. So insofar as the HIF process addresses the last mile problem by improving healthcare systems, this is likely to lessen further such temptations for corruption.

2.2.2 *Question 2: What Risks of Further Corruption Might the Health Impact Fund Bring?*

So far it has been argued that there is a reason to believe that a Health Impact Fund-type scheme could have some lessening effect on corruption in the pharmaceutical sector. But of course, such a schema could potentially generate other ways in which corruption might occur. So we need to explore that possibility: If some sort of scheme like the Health Impact Fund was put into place, what new opportunities for corruption might there be? (And might this mean that the net amount of corruption does not decrease after all?)

The central issue is that the Health Impact Fund's proper functioning requires a reliable and trusted process of assessing the causal impact of the various pharmaceuticals on the global disease burden. Clearly, such a task will be incredibly challenging. From the perspective of the question of whether the Health Impact Find is a feasible idea, we need to ask: Could this even be carried out? How accurate, reliable, or trusted would the results be?

Our question here is more specific: Might corruption previously aimed at manipulation of trial results and clinicians' prescribing behavior just migrate to this process of impact assessment? If so, and if actors such as drug companies are able to bias these assessments, then it won't be so obvious that overall corruption will diminish. Indeed, insofar as the process of impact evaluation is biased or open to corruption, this will affect our argument above that drug companies will only have a motive to deceive and manipulate to get their drug used if the drug will, in fact, have a significant positive impact. This is a very complex matter, which we can only address in outline here.

Assessing whether and how much corruption there will be (and how this compares with present corruption) requires thinking about each of two interrelated matters: methodological challenges and institutional issues. Concerning methodological challenges, we must ask: Given the sorts of things that need to be assessed, and our available methodologies for assessing these, to what extent will the evidence be clear enough to be compelling or uncontroversial? Questions of institutional setup involve such questions as how the process of corruptive influence would go: Who would be corrupted and how? Could safeguards be put into effect that would be adequate here?

Note also that a full analysis would require something *comparative*. We are familiar with the task of assessing safety and efficacy in drug approval. How does this new task of global impact assessment differ from this? The question won't be just whether there would be corruption, but whether there would be more or less. Would such problems be more or less manageable? Would more or less damage occur as a result? Are there differences between the task of assessment of the impact on global disease burden and those such as getting a drug approved or manipulating prescribing habits—in terms of either methods or institutional arrangement—that make the impact assessment more or less open to corruption?

Let us first discuss the matter of methodological challenges. Clearly, the assessment process will be a vast, complex, and difficult undertaking, with potential pitfalls, both methodological and logistical. For in order to put together a causal story, about the real impact of a given drug on the global disease burden, which is adequate to our task and about which we can be confident, there are several kinds of studies to do. Hence the process is complicated, and hence can contain controversial elements.

Epidemiological studies can track the actual change in the incidence of particular diseases. But these methods will not be foolproof as a way of assessing the causal claims, and so it will be necessary to carry out a number of other studies (to confirm various parts of the causal process, and help tease apart various causal claims).

One of these will be (randomized) clinical trials, the usual standard for establishing efficacy for getting FDA approval and for evaluating the rationality of use. It is often noted that what such trials can really assess is potential efficacy. They don't necessarily assess actual impact, because this is something which is affected by a particular population being different from the study sample, as well as by whether the infrastructure of the local healthcare system results in proper usage.

Other studies would include the tracing of random samples of the product to end users, and the statistical analysis of correlations between sales data and variations in the incidence of the target disease. We can measure independently such things as how the drugs are being used, whether they are being adequately stored, and other intermediate causal factors.

No one type of study is adequate for the whole task, but it is plausible that the whole set of studies can provide adequate evidence. But we must also note that any given study involves certain methodological choices and hence can be debated. Recall the ways in which the drug companies have used methodological choices to manipulate study design in a way that biased the case. These decisions about powering and selection of the study population could play a role here in the global disease burden assessment as well. Thus, those overseeing the process will have to make a number of judgments, such as those concerning:

- which studies and how they are designed (as before)
- what outcome measures (including the issue of QALYs (quality-adjusted life-years), since the real benefits include not just reduction of mortality and morbidity, but the quality of life
- generalizability (and thus when a study needs to be repeated in a potentially new context): decisions about whether and to what extent to replicate RCTs in multiple environments (rather than just one overall study)
- how to weight and integrate different sources of information: For all these studies will have to be amalgamated into an overall judgment about what the impact has been.

Of course, the more there is a clear and uncontroversial method for carrying this out, the less room there could be for hidden manipulation. The more there is openness to interpretation on what the criteria are, the more this might allow a place for someone to create an improper influence—and relatedly, less trust in the results.

Finally, there may be some especially tough cases, ones which raise not just empirical questions but also "conceptual" questions of what we mean to measure. For instance, how adequately will we determine the positive impact of a vaccine if we measure it (only) over the first 10 years, given that benefit is expected to last far into the future? As another example, how does one identify the relative contribution of a drug whose contribution depends on other factors? In one scenario, it depends on background nutrition, and we may ask: If we just rely on the actual background context, then a company will just be lucky that nutrition levels were relatively high that year. In another scenario, suppose a drug's impact depends on the presence of some other drug. If this other drug is made by the same company, then perhaps this would not matter, but this will be different if it is made by a competing company.

All of these could generate worries that there often won't be a clear answer and that this weakness can be exploited. There may well be good answers about how to deal with such cases, but we are still left with some understandable skepticism.

Now, to this worry that the process will be very difficult, *one* sort of response is that we are likely to get better at the process over time. This may seem like too glib a response, but I think it should be taken seriously for the following two reasons:

The first refers to an analogous sort of task for medicine in general. The problems such as those of amalgamation come up in other contexts of evaluation in medicine, and they make up the subject matter of the project of "evidence-based medicine". These are difficult, but it is certainly taken seriously that the right response is to continue to improve the process. For instance, judgments concerning the "hierarchy of evidence" are controversial, but plausibly they are workable, and progress is being made.

The second reason for optimism involves the following. The task of gaining this knowledge about the impact on global disease burden, as well as gaining the knowledge about the best methods for assessing this, is certainly difficult. But it should be looked at as not some new (and onerous) project that we now need to carry out due to our reliance on this new mechanism for adequately carrying out this Health Impact Fund-task of distribution monetary reward, such that the difficulty is simply a strike against the feasibility of the Health Impact Fund proposal. Rather, this is knowledge that we—public health professionals, institutions and researchers—need to obtain (and get better at obtaining) anyway. We need this for a wide range of health policy and planning endeavors, and for countering a variety of unwarranted claims in the health domain. Quite independent of the Health Impact Fund project, it is very important that we improve our skills at gathering and assessing that evidence. Thus if HIF provides an opportunity to get better at it, this counts as a reason to take this up.

Let us turn now to some institutional implementation questions: There will, of course, need to be some sort of committee in charge, overseeing the process of making these assessments (and thus making decisions about the various methodological matters). Ideally, their deliberation and oversight can be done in a transparent manner, and with safeguards such that it cannot be easily or effectively manipulated. So, how might pharmaceutical companies influence these assessments? They could bribe members carrying out or in charge of the process (or any of the others involved in the process) to try to get them to overestimate their drug's impact. It doesn't seem that this would be done by giving them pens and mousepads and luxury vacations, but perhaps there are subtler ways in which this could be done. But it does seem that this will work less well than in the case we are familiar with—when the goals are simply to manipulate physicians' prescribing behavior, and where this is done by drug reps striking up relationships with busy and demoralized clinicians. What would require further analysis and comparison would be an inappropriate influence that occurs concerning the FDA or other national regulatory bodies. Here it may be necessary to rely on other methods for curbing corruption.

2.3 Involvement of Drug Companies?

Whatever such a process is put together, a crucial requirement is that it be as *transparent* as possible, but it is also likely to rely on some input from some outside sources. On the face of it, we will also want to make public the data from the various sorts of studies involved. This suggests a process of gathering comments from others, and these others might include the drug companies. And this raises an important question: Much harm has been done by the actions of pharmaceutical companies via the deceptive design of trials, manipulation of beliefs and actions of clinicians, etc. One reason that we might have hope that this HIF-assessment process can be carried out effectively and with minimal corruption might be precisely that the drug companies could be kept out of this process. So surely, it will be said, we should, in fact, set things up so that the drug companies play no role.

Note that this is not simply the "divestiture strategy" mentioned above. It's not about public funding of drug development; drug companies would continue to make decisions about what drugs to develop based on what profit they could make, but the new sort of incentive of the HIF is hoped to motivate them to make what in fact are socially positive decisions about drug development. But this HIF proposal is compatible with keeping drug companies out of the process of about drug evaluation.

But before we accept this position, we should consider the following:

First, there are different levels of involvement. Clearly, drug companies should not have an organizational or leadership role, but this does not mean that they would not be involved at all. So we need to ask: could they provide evidence? Could they do or be involved in carrying out a study?

Note that on one scenario, that oversight committee would be open to all inputs of evidence (some from various governments, some from its own scientists, and also some from drug companies), and then use its own judgment to make an overall assessment. In another scenario, it would eschew all inputs from drug companies. Or perhaps there is something in between: perhaps precise criteria could be specified that any study they will look at must meet; the hope would be that even studies by drug companies would not be able to be distorted if it is clear that the study meets various strict criteria (minimally, for example, the provision of all of the data). (Of course, one might point out that this should be required not just for the HIF track, but in the case of the traditional patent track as well.)

Of course, nothing is foolproof—a drug company could push a certain methodological choice (power of the study, dose and context to be tested, etc.) and give a reasonable-sounding rationale for this, and yet their real reason for doing this might be that it biases the study in their favor. Clearly, the agency would have to be both as disinterested as possible (not subject to corruptive influence) and also methodologically sophisticated.

Note that conceivably there would be the following benefit to drug company involvement. This is that competitor drug companies would also be able to weigh in in this way, so this gives some more hope for diminishing such bias. This is different from the (FDA) "approval" process, which is not comparative. The HIF process is a head-to-head comparative process to see what proportion of the disease burden that was lifted was lifted by one drug rather than another. On the face of it, this is a benefit.

Second, we must decide whether it is feasible to eschew drug companies from the process. We need to ask what the consequences would be of rejecting drug company involvement. Would the drug companies accept such a situation? Do we need their money or resources, or even just their buy-in? One important question here is the following: Might it be argued that in addition to these things, the drug companies would have special expertise such that it would be a significant loss to not have them involved at all?

One might respond to this that, given the kind of information the Health Impact Fund needs to collect (not whether a drug works better than a placebo in a controlled study, but whether a difference in health outcomes has been made out in the world), one wouldn't expect the drug companies to have irreplaceable expertise. After all, this has not been where they have been focusing their attention under the regime of "maximize price and maximum sales." So perhaps the dilemma is not so great and the committee should simply not accept data or studies from drug companies.

But, to probe a bit further, note the following: First, given the reorienting of thought brought on by the Health Impact Fund project (especially keeping in mind the "last mile" problems), the drug companies are likely to get better at this. And second, we would *want* them to get better at this, because it's *good* if they reorient their thinking about what's involved in making a positive contribution to global health in ways that take these things into account. So the decision to keep them out of the loop may not be so obvious.

The issue remains complicated: Given their previous behavior, the risks of leaving them in the loop may be too high. But this line of thought does raise the issue of whether the drug companies would accept such a situation, as well as whether this endeavor needs their money or resources.

2.4 Conclusion

The Health Impact Fund is an interesting but controversial proposal that deserves further exploration, including its connection to the topic of corruption. Its explicit purpose is to motivate the development of drugs for the global poor and make them available at an affordable price. But reasons have been given here for the claim that Health Impact Fund would have some dampening effect on various sorts of corruption, so these could count as further reasons in its favor. Further, note that this fact could be studied to see if it might provide other insights for methods of constraining corruption.

But it also generates some further contexts for potential corruption, and so serious attention must be paid to this going forward. And we need to see how best to control for or put in check or minimize the potential corruption and bias in such assessment.

References

Brody, H. (2007). *Hooked: Ethics, the medical profession and the pharmaceutical industry*. Lanham, Maryland: Roman and Littlefield.

Crombie, J. (2020). Medical ghost- and guest-writing as corrupt practices and how to prevent them. In K. T. Çalıyurt (Ed.), *Integrity, transparency and corruption in healthcare & research on health* (Vol. 1, pp. 141–158). Singapore: Springer.

Eyal, N. (2012). Global health-impact labels. In Millim & Emanuel (Eds.), *Global justice and bioethics* (pp. 241–278). Oxford University Press.
Hollis, A., & Pogge, T. (2008). The health impact fund: Making new medicines accessible for all (Incentives for Global Health, 2008). http://healthimpactfund.org.

Fred Gifford is Professor of Philosophy at Michigan State University, East Lansing, Michigan, USA.

Chapter 3
Outsourced Services in Private Healthcare Organizations and ISAE 3402 Assurance Engagement: A Research Focused on Turkey

Esra Atabay and Engin Dinç

Abstract Outsourced clinical and nonclinical services have become widespread since the beginning of the Healthcare Transformation Project in the 2000s in Turkey. The objective of ensuring efficient use of resources requires the control and auditing of outsourced clinical and nonclinical services in healthcare organizations. However, there has been no study performed on the effectiveness of inspections and audits of outsourcing in healthcare organizations. Therefore, the control and auditing of outsourced services in healthcare organizations constitutes the subject of this study. The purpose of the study is to assess whether the ISAE 3402 standard is applicable for the control and auditing of support services purchased in private healthcare organizations and to explain through a model how the ISAE 3402 standard will be implemented. To achieve the objective of the study, a survey was conducted on the managers of 25 private healthcare organizations that are included in *a Private Healthcare Group with a Hospital Chain*. (The private healthcare group did not give consent to the disclosure of its title.) The data obtained were analyzed by using cross-examination, single and multiple frequency analysis, and descriptive statistics. As a result of the analysis of the data obtained from the study, it was revealed that nearly half of the private hospital managers who participated in the study had heard about the ISAE 3402 standard and the vast majority believed that the assurance engagement conforming to this standard was feasible in the private healthcare industry. Moreover, it was explained via a model how the ISAE 3402 Standard would be applicable.

Keywords Outsourcing · ISAE 3402 · SSAE 16 · Private healthcare organizations

E. Atabay (✉) · E. Dinç
Trabzon University, Trabzon, Turkey
e-mail: eatabay81@gmail.com

E. Dinç
e-mail: engindinc66@gmail.com

K. T. Çalıyurt (ed.), *Integrity, Transparency and Corruption in Healthcare & Research on Health, Volume I*, Accounting, Finance, Sustainability, Governance & Fraud: Theory and Application, https://doi.org/10.1007/978-981-15-1424-1_3

3.1 Introduction

Organizations must adapt to the increasing competition and changing technological conditions so as to improve profitability and productivity further. Adaptation to such changing conditions also requires bearing excessive costs. To minimize costs, organizations have started to outsource services that have the status of subsidiary activities.

Today, healthcare organizations outsource some part of clinical and nonclinical services. For healthcare organizations, it is essential to assure third parties to outsourced clinical and nonclinical services. Such assurance provided for third parties in healthcare organizations is highly effective in creating client satisfaction and loyalty. The loyalty and satisfaction created a positive contribution to the profitability of the business indirectly.

The concept of support service and support services outsourced in private healthcare organizations are addressed in this study. The study addresses each topic individually, including what services are outsourced to private healthcare organizations, for which reasons outsourcing is preferred, what kind of benefits and disadvantages outsourcing brings to the organization. What type of audits are carried out in healthcare organizations and by whom and by which criteria the outsourced services are audited? It is also noted in the study that an assurance engagement conforming to the ISAE 3402 Standard regarding Assurance Reports on Controls at Service Organizations can be used as a basis for the auditing of services outsourced in private healthcare organizations in Turkey, and the relevant control model is described.

The ISAE 3402 assurance engagement, on which emphasis is laid in the study, is performed by independent auditor firms, called as the Big Four in Turkey, for a specific price (nearly 50–100 thousand TRY) upon request of the management, without any legal obligation to that effect. As public healthcare organizations do not seek profitability, private healthcare organizations were preferably included in the scope of the study.

The final part of the study, which is based on research, includes findings obtained through an analysis of the surveys answered by the managers of a Private Healthcare Group having a leading status in the private healthcare market in Turkey and having the highest number of hospitals across the country. The frequency analysis performed and the descriptive statistics are generally aimed at revealing what kind of support services are outsourced by private healthcare organizations in Turkey, by whom such outsourced functions are controlled and audited and for what purposes such control and auditing are performed, to what extent there is an awareness of the ISAE 3402 standard and trends for procuring an audit conforming to the mentioned standard and their causes. Furthermore, it is described through a model how the ISAE 3402 standard is explicitly implemented to a support service.

3.2 An Overview of the Concept of Outsourced Services

In its most straightforward sense, Support Service is defined as the outsourcing of a product or service produced by a company (Beaumont & Sohal, 2004: 688). The concepts of outsourcing and support services are used in the same sense. Today, most companies tend to outsource services as they intend to focus their principal activities by performing their operations in a more efficient, productive, and profitable manner (Atabay, 2016: 50). Support services can be defined as services in which works that are of secondary significance are shifted, and external suppliers assume the activities (Onaydın Yıldırım, 2011: 37; Kalkan et al., 2015: 38). The use of support services or outsourcing requires the signing of a contract for a third party to provide goods and services for the host organization (Foxx, et al., 2009: 42).

Technological developments have been making quite fast progress in today's world. For organizations alone, it is hard to keep up with that progress, particularly regarding meeting the increasing costs. Organizations procure support services both to maintain their prices at a low level and to adapt themselves to technological developments. In the study conducted by Bayındır (2007), the factors leading organizations to choose to outsource within an intensely competitive environment are addressed under the topics of organizational factors, cost-based factors, financial factors, human resources, quality enhancement, and technological innovations. In their studies, Gözüküçük and Çelik (2012) examined the reasons which direct organizations to procure support services in two categories; as internal and external factors. They listed the internal factors as reducing costs, increasing flexibility, downsizing, ensuring cash flow, focusing on necessary skills, lowering risks, redistributing resources, creating a vision, reducing investment expenditures, and improving quality. And external factors as maintaining contact with competitors, responding to client expectations, keeping up with technological developments and opening up to new markets.

Organizations procure support service to gain many advantages. However, the procurement of support service may lead to disadvantages while providing benefits for the organization. Such powers and problems can be classified as strategic and tactical, long-term and short-term, concrete or abstract (Beaumont and Sohal, 2004: 691). Outsourcing provides organizations with the opportunity to

***control and reduce operational costs,
***focus on core activities and competencies,
***respond to variable and volatile demands,
***avoid cultural problems,
***reduce costs and technology risks,
***increase flexibility,
***convert fixed costs to variable costs,
***ensure lower capital dependence,
***improve the quality of activity,
***specialize, assuring standardization of service,
***gather motivation and employee skills together,

***reduce the risks caused by ambiguities in the market, and
***perform a rapid growth. (Bailey et al., 2002: 84; Guimarães and Carvalho, 2011: 141; Kalkan et al., 2015: 38; Döğücü and Sayım, 2009: 265; Yalçın et al., 2011: 85).

Without a doubt, the procurement of support services (outsourcing) may cause significant problems within the organization despite the aforementioned potential benefits. Such issues may include the risk of losing control of the activities, the risk of being exploited by suppliers (Bailey et al., 2002: 84). The potential of suffering high incremental costs, disobedience of suppliers and degradation of quality, emergence of follow-up costs which did not exist previously, emergence of cultural differences caused by internationality, the possibility of dissatisfaction with the support service received, lower impact than expected on costs, challenges in integration and coordination (Guimarães and Carvalho, 2011: 141). The loss of long-term research and development competitiveness and the service supplier's loss of required competence (Koszewska, 2004: 230).

When deciding upon the support services to be procured, organizations perform a cost–benefit analysis to consider whether internal supply or outsourcing of the activity will be more profitable for the organization. Organizations usually tend to outsource such operations that remain out of their primary business and require high investments. Events such as information technologies, parts production, distribution, finance, accounting, real estate management, human resources, telecommunication, manufacturing, call centers, field sales, marketing, building management, and transportation are examples of support services (Nazlıoğlu and Yar, 2016: 73). Support service procurements, which were initially efficiently used in metal production business in the nineteenth century, were then observed in fields such as cleaning, security, catering industries in the 1960s and later. In line with the technological developments in the 1980s, in the form of outsourcing of the management of specific activities that are not related to core competencies of organizations (Yalçın et al., 2011: 84).

3.3 Support Service Procurements in Healthcare Industry, Reasons, Benefits and Risks

Although the outsourcing practice in healthcare organizations in our country started with limitation to nonmedical services of hospitals such as the domestic economy, security, and catering services. It has recently become a general practice also covering administrative activities such as invoicing, data recording, etc. to adapt to environmental and economic changes (Ministry of Health, Republic of Turkey, 2010). Like many other public and private organizations, healthcare organizations prefer enterprises specialized in their field so that they will focus on their primary activities and provide savings on the cost of investment required for numerous subsidiary activities. These activities can be listed as such as security, dining service, cleaning, paging service, information and software, invoicing, data recording, etc. Such outsourced services are referred to as support services.

Today, healthcare organizations serve their clients by receiving a significant portion of clinical and nonclinical services from organizations specialized in their respective field.

Upon a literature review, it is observed that there is a large number of factors directing healthcare organizations to outsource services. According to the result of the research carried out by Döğücü and Sayım (2009), healthcare organizations procure support services for the following purposes, respectively: to increase service quality, to improve patient satisfaction, and to ensure that innovations are promoted. The same study examined the reasons why healthcare organizations do not refer to the support service procurements, and those reasons were identified as follows, according to their respective order of priority: the applicable legislation, administrative decisions, and the opinion that internal activities are performed in better quality (Döğücü and Sayım, 2009: 271). Healthcare organizations refer to outsourcing intending to reduce costs of labor, share potential risks, keep up with the technology, enhance quality and productivity, and give flexibility to the organization (Sezer, 2009: 36). Upon the research conducted by Gözüküçük and Çelik (2012), it was found out that the reasons leading healthcare organizations to procure support services include the opportunity of easier access to technology, less time, money, and resources for maintenance-repair and improvement of medical devices, and shorter wait times for doctor's appointments. The reasons for healthcare organizations to procure support services are listed as ensuring cost savings and increased productivity and current quality according to the study performed by Karahan (2009), while the study by Yiğit (2004) lists the reasons as follows: reducing costs and sharing risks by improving service quality and increasing flexibility. Managers of healthcare organizations seek innovative methods that will quickly and efficiently provide the added value they need for services, improve the productivity of the operations performed by the organizations, and improve harmony with the regulations (Chen and Perry, 2004: 24). In line with the information obtained from the literature, the fundamental factors leading managers to refer to support service procurement can be listed as follows: ensuring cost savings, improving client relations and performance, and having the opportunity to share information and better control the services (Sarivougioukas and Vagelatos, 2002: 58). Upon their research carried out in five different countries, Guimarães and Carvalho (2011) identified the reasons for support service procurement in the healthcare industry as follows: ensuring lower human resources cost, lower capital expenditures and smaller investments in equipment; standardizing service quality; enabling cost reduction in subsidiary activities; allowing redesign of business processes, development of information technologies, access to expertise, flexibility, focus on critical activities, prevention of conflicts between public and private organizations; reducing the need for specialized staff; providing risk migration and patient satisfaction.

Healthcare organizations receive support services for the reasons above, and such support services procured provide significant benefits for hospitals. Upon outsourcing of a subsidiary activity that can be provided within the organization, healthcare organizations possess the latest technologies, quickly respond to requirements and environmental changes, provide employee satisfaction, boost morale—motivation and productivity, increase efficiency and quality, build a sound public audit mechanism, with reduced burden in subsidiary activities, specialization in medical services, time and cost savings, sharing of risks and higher adaptability to patients' requests and to developments. (Çakırer, 2003: 217–218). According to Nazlıoğlu and Yar (2016), support service procurements not only provide considerable savings in costs of healthcare services but also increase productivity. In his doctoral dissertation, Akyürek (2013) researched the use of outsourcing in the healthcare system after that he found that the most common opinion of managers of healthcare organizations about benefits of support services received is that it increases the quality of service. The healthcare organization's outsourcing of services has increased patient satisfaction, in addition to enhancing the quality of service, and ensured that the requirement of an efficient function is fulfilled. According to the opinion of the managers, support service procurements contribute, at a minimum, to the increase in flexibility of the organization (Akyürek, 2013: 116). There has been a large number of studies conducted on contracted service suppliers in Canada, the United Kingdom, and New Zealand, the primary results of which include the facts that support service procurement improved performance, provided cost-saving, and increased management time in main operating activities (Moschuris and Kondylis, 2006: 5).

In addition to obtaining benefits by outsourcing some services, healthcare organizations also experience many problems due to such services. Upon research performed, it has been determined that outsourcing service or product reduces quality, causes delays in supply and leads to difficulties in control (Döğücü and Sayım, 2009: 274). Support service procurement requires having an excellent management team. Unless the support services procured are adequately and appropriately managed, they may become a significant problem for the organization rather than providing benefits. Therefore, healthcare organizations are required to employ managers with strong leadership qualifications and the ability to manage outsourced services, which means an increase in personnel costs. Losing control of outsourced services, healthcare organizations may become overly dependent on the supplier company. Moreover, the healthcare organization may need to take precautions against the possibility of disclosure of patient information by the supplier company (Sezer, 2009: 38–40).

Throughout many years, healthcare organizations have been making use of support service activities such as cleaning, catering, laundry, collection, clinical/diagnostic material maintenance, foreign language services, emergency unit, anesthesia, hospital call center, cardiology, medication, security, information systems, operating activities, indoor parking, rehabilitation, and patient records all over the world (Miller and Washington, 2007, as cited by Akbulut et al., 2012: 27). The research study carried out by Moschuris and Kondylis (2006) revealed that the support services most frequently used by healthcare organizations include cleaning, security, cafeteria, legal, clinical/diagnostic material maintenance, information systems, laundry, laboratory, and catering services, respectively. The outsourcing rate has increased since the onset of the Healthcare Transformation Project in the 2000s in Turkey. Healthcare organizations have tended to outsource nonclinical services such as cleaning, automation, security, dining service, laundry, ambulance, patient help desk, and cafeteria, as well as clinical services such as magnetic resonance (MR), computed tomography (CT), laboratory, and ultrasound (USG). It was observed in a study conducted in Turkey in 2009 that the three support services most commonly procured by healthcare organizations, respectively, include catering, cleaning, and security services, while laboratory services constitute the least widely acquired service (Döğücü and Sayım, 2009: 270). In another research carried out within the same year, it was seen that cleaning, computer automation, and security services are the most commonly outsourced services, and radiology services are the least widely outsourced service. (Karahan, 2009:195). According to the report of the research conducted by Refik Saydam School of Public Health, Ministry of Health, the Republic of Turkey in 2010, the top three nonclinical services outsourced by the healthcare industry in Turkey are cleaning, automation, and security, and the top three clinical services outsourced are magnetic resonance, computed tomography, and biochemistry laboratory, respectively (Ministry of Health, Republic of Turkey, 2010: 28, 30).

3.4 Auditing of Support Service Procurements in Private Healthcare Organizations

Although healthcare organizations had been subject to audits under the title of a general examination by ministries of health until the year 2008, a new audit model has been in use primarily at hospitals under the title of "regularity audit" since 2008 (Private Hospitals Audit Guideline, 2014: Introduction). Regularity audit is defined as disclosure of opinion by public administrations on reliability and accuracy of their accounts and transactions as well as financial reports and statements; determination of whether or not the accounts and operations of public administrations regarding their incomes, expenses and assets are in conformity with the laws and other legal arrangements, and the assessment of financial management and internal control systems (Regularity Audit Guideline, General Framework, 2014: 3). While the definition of regularity audit addresses an audit model related to accounts and transactions of public administrations, the Ministry of Health has prepared audit guidelines individually for each unit such as private healthcare organizations, oral and dental health hospitals, etc., by taking this audit model as basis, and published the guides on the website of the Ministry of Health of the Republic of Turkey. According to this audit guideline prepared for private healthcare organizations, all units within the healthcare organization are audited regarding their conformity with the Private Hospital Regulation. In other words, the Ministry of Health performs an on-site audit to find out whether the unit possesses the necessary devices, equipment, and personnel as stipulated in the Regulation and whether it meets the technical criteria required (surface area, number of doors and windows, door swing direction, etc., …) for the launch of the unit within the healthcare organization.

Audits in private healthcare organizations are grouped into two main categories depending on their purpose and basis. Checks can be categorized as license audits, service audits, rating and registration audits regarding their goals; and as ordinary, extraordinary, and periodic inspections regarding their basis. Licence audit refers to an examination which is carried out at least once a year by a team of auditors, including at least the technical personnel, to identify whether the private healthcare institutions and organizations licensed by the Ministry of Health as per the relevant legislation fulfill the licence requirements, depending on the quality of the licence and the criteria indicated in the audit form. Service audit means an on-site inspection by which the presentation of services by private healthcare institutions and organizations is assessed at least biannually by a team of auditors regarding conformity with the legislation and arrangements of the Ministry. Rating and registration audit means an on-site review routinely carried out at least biennially in a frame of the relevant law during the rating and registration process of qualified units. Routine inspections performed at private healthcare organizations as aforementioned within specified periods are referred to regular audits; while audits held upon any notice, complaint or sua sponte are applied to as extraordinary examinations. Periodic reviews mean such checks that are performed at the end of a specified period granted for the elimination of deficiencies identified during both ordinary and exceptional audits.[1] The audit team referred to in the directive means a group of at least 2 auditors including one physician, appointed by the provincial directorate of health to perform audits.

Audits are performed at private healthcare organizations by quality units and audit teams of hospitals, the Ministry or international accreditation institutions for reasons such as improving service quality and patient satisfaction at individual healthcare organizations and ensuring continuous improvement of performance. Both international quality standards (ISO 9001:2000 Quality Standards) and JCI Accreditation Standards can be used for such audits that are performed for the purposes mentioned above. JCI audits guarantee assessment and certification of whether a healthcare organization has fulfilled a series of standards designed to improve the quality of patient care, to provide a safe environment and to reduce risks toward patients/employees (http://www.uhbd.org/PDF/sunumlar/JCI_ISO.pdf). Such audit is not of mandatory nature; healthcare organizations in the scope of accreditation are subject to JCI audits.

Beginning from the year 2003 in which there was an increase in outsourcing practices in private healthcare organizations, details about whether the services outsourced by the healthcare organization are subject to on-site audits and whether the records of such reviews are kept are specifically inquired (Türkeli, 2011). Furthermore, it is described in various articles of the Private Healthcare Organizations Service Audit Guideline what kind of physical requirements must be fulfilled by the healthcare organization if the services are outsourced; however, no explanation is provided about their audits (Regulation on Private Healthcare Organizations, Art. 36). The control of such kind of subcontracted services is usually conducted by audit teams formed within the healthcare organization, or quality units or an audit team assigned by the Ministry. Whereas, the satisfaction of patients and patient relatives,

[1] Articles 12–17 in Chapter Four of the Directive dated 05.08.2014 and numbered 1449 on the Procedures and Principles of Audits on Private Healthcare Institutions and Organizations by Provincial Directorates of Health.

confidentiality of information and responsiveness to demands are quite significant especially for private healthcare organizations. Such requirements can be fulfilled through controls to be performed by an external audit team including technical personnel specialized in their field. In this frame, assurance engagement can be realized in compliance with the assurance reports standard regarding controls in support service organizations, which is known as ISAE 3402 internationally and as GDS 3402 in Turkey. Commonly used to assure third parties especially about outsourced information systems, this standard is a preferable audit model both for information systems and any other outsourced support services of healthcare organizations.

3.5 An Audit Model for the Control of Support Services in Private Healthcare Organizations: ISAE 3402 Assurance Engagement System

The assurance engagement conforming to the ISAE 3402- Assurance Reports Standard Regarding Controls in Support Service Organizations is an assurance audit standard preferred by healthcare organizations outsourcing services for the audits of such facilities. The mentioned assurance engagement is not a mandatory audit, but it refers to an assurance provided by independent experts for beneficiaries of the service of the services offered by the organization. A check is performed upon request by the organization outsourcing a function. This indicates that the management of the organization is willing to incur the costs regarding the execution of the assurance engagement. Therefore, assurance engagements conforming to the ISAE 3402 standard constitute a new assurance audit model that can be recommended especially for healthcare organizations in the form of private hospitals rather than public hospitals.

This model is built upon a tripartite relationship. The interrelated parties are the organizations providing support service (service organizations), organizations outsourcing support service (user entity), and audit firms or individual auditors offering assurance service. Assurance engagement may either be demanded by the healthcare organization outsourcing service (procuring support service) or by the firm that will provide requested support service.

The mentioned assurance engagement process is modeled and described below:

The "Private hospital" receiving support service may refer to the audit firm to demand assurance regarding controls of the support service procured, or the company providing support service may forward to the audit firm a request for the audit to offer assurance for the parties about the service it provides. The audit firm assesses the received request for assurance and carries out assurance activities. The scope of such activities varies depending on whether the "organization providing support service" has any subcontractor organization. In case the organization providing support service has a subcontractor organization which is also required to remain in the scope of an audit, the audit firm will include the subcontractor organization in the scope of assurance activities along with the "organization providing support service." Such type of assurance engagement is referred to as the general method. If the "organization providing support service" is not required to remain in the scope of assurance controls despite having a subcontractor organization, or if there is no subcontractor organization, then the audit firm carries out assurance activities by using the method called Narrowed Method (Atabay, 2016: 110).

This study is aimed at ensuring that the assurance engagement in conformity with the ISAE 3402 standard is comprehended and implemented by private healthcare organizations procuring support service. Accordingly, the audit model in which the request for audit is fulfilled by the private healthcare organization receiving support service is shown in Figs. 3.1 and 3.2.

Figure 3.1 shows a request by a private healthcare organization outsourcing service, for assurance engagement by an independent audit firm to assure the design, implementation, and operation activities of controls related to the services outsourced. According to the model, the organization providing support service does not have any subcontractor organization. Thus, assurance engagement covers only the

Fig. 3.1 Example 1—To assurance services model (narrowed model). *Source* Atabay (2016: 111)

Fig. 3.2 Example 2—To assurance services model (comphrehensive model). *Source* Atabay (2016: 112)

organization providing the relevant support service (Narrowed method). This model involves the signing of a written service contract between the private healthcare organization and the firm from which the support service is procured. To assuring third parties regarding such service purchased, the private healthcare organization forwards a request to an independent audit firm for an assurance engagement in conformity with the ISAE 3402. Thus, an assurance service contract must be signed between the audit firm and the healthcare organization. The audit firm performs audit activities by auditing the internal control structure and workflow of the organization providing support service. Depending on the request submitted to the auditor, the auditor issues an assurance report in Type 1 or Type 2 and sends the report to the private healthcare organization.

Figure 3.2 shows that the assurance audit will be performed as per the comprehensive method. Unlike Fig. 3.1, the audit firm is responsible, in addition to the review of the "organization providing support service" as required, for auditing the subcontractor organization of the firm under the same conditions

About the services outsourced, private healthcare organizations increase the satisfaction of patients and patient relatives by providing them with a high level of assurance with the assurance report obtained upon audits performed by independent auditors in conformity with the ISAE 3402.

3.6 A Research Focused on Support Service Procurements and Audits of Private Healthcare Organizations

Research Methodology

Purpose of the research: The goal of this research is to reveal the applicability of the ISAE 3402 standard in auditing of all services outsourced in the private healthcare industry and to identify the opinions of relevant hospital managers about this standard. Three survey questions were selected for this purpose. The survey questions are as follows:

1. What are the support services outsourced to the private healthcare industry?
2. How are these services controlled and audited?
3. What are the opinions of hospital management about the applicability of the ISAE 3402?

Scope and Limitations of the Research: This study focuses solely on the private healthcare industry in the Turkish healthcare sector. The services related to the ISAE 3402 assurance audit referred to in the survey are not mandatory in Turkey; independent audit firms provide them called the big four (PWC–E&Y–D&T–KPMG) upon request by the hospital management. As Turkish public healthcare organizations do not seek profitability, private rather than public healthcare organizations were included in the scope of the study. The research was carried out in hospitals of a *Private Healthcare Group*, which is spread across Turkey with the highest number of hospitals in the country. The study could not be performed in all private healthcare organizations due to time and cost limitations.

Sample Group and Data Collection Method of the Research: The sample group of the research consists of General Managers, Deputy General Managers, Hotel Management and Support Services Managers, Managers of Administrative Affairs and Purchase Managers of 25 private hospitals owned by a *private healthcare group* and operating in various regions of Turkey. The survey method is used in the research as the data collection method. The survey form consists of four parts; the first part includes questions about demographic features of the participants, the second part contains questions about types of support services procured by the hospitals, the third part contains questions about the control and auditing of support services obtained, and the final section includes questions about awareness of the hospital managers on the ISAE 3402 and the applicability of the standard at the hospitals. The survey form prepared was sent via e-mail to nearly 100 managers primarily including purchase managers, managers of administrative affairs, and hotel management and support services managers. 52 replies were collected in total from managers of the 25 hospitals, including minimum (1) from each of the hospitals. 49 data are considered adequate for significant results at a significance level of 95% and a sampling error of 10% (https://www.surveysystem.com/sscalc.htm).

Validity and Reliability of the Research: Validity means the extent to which a test or a scale is measuring what it purports to measure. It is of importance to determine whether the survey form prepared contains statements in conformity with the purpose of the study. Relevant academicians tested the content validity of the obtained data. The reliability of the scoring comments in the final part of the survey form used in the research was measured by using the Cronbach's alpha coefficient. The calculation was made by using SPSS 23 Statistics software, according to which the reliability coefficient related to the scoring performed by the participants in two separate sections was found as 0.751 for 7 statements and 0.785 for 9 comments. Such reliability levels are indicated as "highly reliable" (Kalaycı, 2014: 405).

3.6.1 Research Findings

Findings Related to Demographic Features of Participants

Frequency values regarding demographic features of 52 participants taking part in the research are presented in Table 3.1.

According to Table 3.1, over half of the participants taking part in the research include the managers of administrative affairs and purchase managers, and over half of the participants have an undergraduate degree. The participants who answered the survey form are mainly between the ages of 31 and 40. 59.6% of the participants have management experience of 1–10 years, while 32.7% have management experience of 11–20 years. 57.6% of the managers reported that they acquired the necessary knowledge to be qualified as managers during their undergraduate education, while 24.2% stated that they were trained as managers of the in-service training

Table 3.1 Findings related to demographic features of participants

		n	%			*n*	%
Title	General Dir.	4	7.7	Educational status	Associate degree	6	11.5
	Dep. General Dir.	10	19.2		Bachelor's degree	31	59.7
	Hotel management and support services Manager	9	17.3		Master's degree	15	28.8
	Dir. of administrative services	16	30.8		Doctoral Degree (Ph.D.)	–	–
	Purchase Mngr.	13	25.0				
Gender	Male	31	59.6	Age	30 and younger	6	11.5
	Female	21	40.4		31–35	16	30.8
Management experience	Less than 1 year	1	1.9		36–40	16	30.8
	1–5 years	16	30.8		41–45	6	11.5
	6–10 years	15	28.8		46–50	3	5.8
	11–15 years	9	17.3		Over 51	5	9.6
	16–20 years	8	15.4				
	21 years and longer	3	5.8				
Management experience in the organization	Less than 1 year	4	7.7	The region in which the hospital is located	Black Sea region	8	15.4
	1–5 Years	25	48.1		Aegean region	3	5.8
	6–10 Years	18	34.6		Mediterranean region	1	1.9
	11–15 Years	5	9.6		Marmara region	33	63.5
	16–20 Years	–	–		Central Anatolian	6	11.5
	21 years and longer	–	–		Eastern Anatolia	1	1.9
Type of qualification as manager	By undergraduate degree	38	57.6				
	By postgraduate degree	8	12.1				

(continued)

Table 3.1 (continued)

		n	%			*n*	%
	Upon certification training given by a private organization	4	6.1				
	Through in-service training	16	24.2				

they received. When the regions of the hospitals in which the managers work are analyzed, it is observed that managers from all areas mainly including the Marmara region answered the survey form.

Findings on Support Service Procurements of Private Hospitals
The second part of the survey form sought answers to the question of whether the private hospitals outsource services, and if yes, which clinical/nonclinical services they outsource. The following findings were obtained through the multiple frequency analysis performed.

According to Table 3.2, all survey participants outsource services. As for clinical services, pathology services constitute the service outsourced at the highest rate (20%) among the services outsourced by the *private hospitals of the Private Healthcare Group* included in the research, which is followed by MR, Beauty center, Biochemistry laboratory, and nuclear medicine equally with the second-highest rate (8.9%) and microbiology laboratory service with the third-highest rate (7.4%). The primary nonclinical services usually outsourced by managers as support services include dining service (12.3%), cleaning (11.7%), cafeteria and security (10.7%), pest control and laundry service (9.6%), respectively.

Findings Related to Control and Auditing of Support Services
In the third part of the survey form, the participants were asked whether the support services procured are controlled and audited, and if yes, for what basis they are monitored and reviewed and what subjects are subject to controls and audits, and by whom, how often and by which methods the inspections and checks are performed. The findings obtained from the multiple frequency analysis conducted are summarized in Tables 3.3 and 3.4.

Table 3.3 shows that the support services procured are controlled and audited in all of the private hospitals which took part in the research. Findings on the basis for such control and auditing, parties who conducted the inspections and audits, the subjects of powers and checks, frequency and methods of reviews are detailed in Table 3.4.

Table 3.4 shows that the top-rated basis for control and auditing of support services among the options submitted to the participants is "*Checking of whether the organization providing the service is competent to provide service and compliant with the*

Table 3.2 Findings on support service procurements of private hospitals

		N	%			*n*	%
Are you outsourcing services?	Yes No	52 –	100 –				
Clinical services received	MR	17	8.9	Nonclinical services received	Cleaning	43	11.7
	Beauty center	17	8.9		Invoicing	2	0.5
	Biochemistry lab	17	8.9		Automation	5	1.4
	Medical services	4	2.1		Cafeteria	39	10.7
	Microbiology lab	14	7.4		Security	39	10.7
	Mammography	6	3.2		Tailor shop	8	2.2
	Radiology	14	7.4		**Dining service**	**45**	**12.3**
	Bone densitometer	4	2.1		Personnel service	20	5.5
	Nuclear medicine	17	8.9		Secretary service	3	0.8
	USG	6	3.2		Car park	26	7.1
	Pharmacy	4	2.1		Pest control	35	9.6
	Dental unit	12	6.3		Hemodialysis patient service	3	0.8
	Nursing	4	2.1		Technical maintenance	6	1.6
	Pathology	**38**	**20.0**		Ambulance	26	7.1
	Doppler	2	1.1		Laundry	35	9.6
	Hemodialysis	3	1.6		Patient help desk	4	1.1
	CT	11	5.8		Medical equipment maintenance repair	5	1.4
					Central heating	1	0.3
					Landscape gardening	13	3.6
					Other	5	1.4

Table 3.3 Findings related to control of support services

		N	%
Are the support services controlled and audited?	Yes No	52 –	100 –

terms of agreement" (19.8%). It is followed by "*Checking of whether the outsourced service is performed in a timely and safe manner*" (18.6%) and "*Checking of whether it increases the satisfaction of the hospital*"/"*Checking of whether the outsourced service created a positive effect on the profitability of the hospital*" (17.3%). The support services procured by the options as mentioned above are controlled and audited by "quality units of the hospital (29.4%)", "hospital management (23.5%)", and "Internal control manager of the hospital (14.4%)". It is mostly observed that the teams in the hospital perform the audits. Such teams perform the audits and controls mostly in the form of "periodic controls (30.6%)". Such controls are followed by "quality audits (25.6%)". Unscheduled controls rank the third place with a rate of 18.8%. Inspections and audits for the support services are most commonly performed in the form of *routine monthly controls* (25.6%), through physical examination method (28.6%), through patient surveys (21.1%), and on-site observation (20.5%).

Findings on the Opinions of Private Hospital Managers about the ISAE 3402
In the final part of the survey form, private hospital managers were asked if they have heard about the standard along with the explanatory model regarding the functioning of the ISAE 3402 standard, what they thought about its applicability, whether they previously procured any service on basis of this standard and whether they would like to procure on basis of this standard. In addition, the participants were provided with statements and asked to score the statements between 0 and 5 in order to identify the reasons for their decision to procure or not to procure such service. The findings obtained upon the frequency analyses and descriptive statistical analyses performed are summarized in Tables 3.5, 3.6 and 3.7.

Table 3.5 shows the general opinions of the participating private hospital managers about the ISAE 3402 standard. 55.8% of the hospital managers reported that they have not heard about the rule. The participants who stated that they have heard about the measure were subject to cross-examination to find out their positions, after that, it was found that purchase managers are much more familiar with the subject. 65.4% of the participants were found out to think that the mentioned standard applies to private healthcare organizations. While the participants who previously procured assurance engagement service by the ISAE 3402 standard were at the rate of 5.8%, the hospital managers who intend to purchase assurance engagement service are at the rate of 55.8%. Findings obtained on the descriptive statistics regarding the reasons for the managers' decision to procure or not to procure such service are shown in Tables 3.6 and 3.7.

According to Table 3.6, the most critical reason for hospital managers to intend to request for assurance engagement about the support services they outsource by the ISAE 3402 is their opinion that "The quality of service received will be increased"

Table 3.4 Findings related to control and audit of support services

The basis for control and auditing of support services	*n**	%
Checking whether the internal control environment of the hospital is adequate for service procurement	27	11.4
Checking whether the outsourced function is performed in a timely and safe manner	44	18.6
Checking of whether the outsourced service conforms with the quality standards of the hospital and the accreditation standards	41	17.3
Checking whether it increases the satisfaction of the hospital	41	17.3
Checking whether the outsourced service created a positive effect on the profitability of the hospital	37	15.6
Checking whether the organization providing the service is competent to assist and complies with the terms of an agreement	47	19.8
Parties performing control and auditing of support services	*n**	%
Hospital management	36	23.5
Internal control manager of the hospital	22	14.4
The internal auditor of the hospital	13	8.5
Department of Auditing Services, Ministry of Health	8	5.2
Officials of the provincial directorate of health	10	6.5
Accounting Manager of the hospital	2	1.3
Quality department of the hospital	45	29.4
Accreditation body	5	3.3
Ministry/Directorate of Environment	8	5.2
Independent auditing firms	2	1.3
Other: External Lab. Integration Lab. Officer, Infection Control Committee	2	1.3
Reasons for control and auditing of support services	*n**	%
Upon complaint	17	10.6
Routine satisfaction controls	23	14.4
Periodic commands	49	30.6
Unscheduled controls	30	18.8
Quality audits	41	25.6
Frequency of control and auditing of support services	*n**	%
Daily	19	15.2
Weekly	20	16.0
Monthly	32	25.6
Quarterly	21	16.8
Semi-annually	17	13.6
Annually	16	12.8
Control and auditing techniques for support services	*n**	%
Through patient inquiry method	14	8.7

(continued)

Table 3.4 (continued)

Through patient surveys	34	21.1
Through observation method	33	20.5
Through physical examination method	46	28.6
Through document inspection method	30	18.6
Using mystery inspectors acting like a client	4	2.5

*As multiple options were selected, n seems higher than the sample size ($n = 52$)

Table 3.5 General findings on the ISAE 3402 Standard

		n	%
Have you heard about the ISAE 3402 Standard?	Yes	23	44.2
	No	29	55.8
Do you think that the ISAE 3402 Standard is applicable?	Yes	34	65.4
	No	18	34.6
Have you ever procured assurance engagement service in conformity with the ISAE 3402?	Yes	3	5.8
	No	49	94.2
Would you procure assurance engagement service in conformity with the ISAE 3402?	Yes	29	55.8
	No	23	44.2

Table 3.6 Findings on the reasons for managers to prefer assurance engagement service in conformity with the ISAE 3402

Cause	Total score	Average score	Effect rating
It will reduce costs per unit	95	3.28	7
The quality of service received will be increased	129	4.45	**1**
Patient satisfaction will be improved	127	4.38	**3**
It will have a positive effect on corporate identity	128	4.41	**2**
Third parties will be offered high assurance	122	4.21	4
The hospital will receive a higher demand	110	3.79	6
It will have a positive effect on the profitability of the hospital	112	3.86	5

(mean: 4.45). This is followed by the opinions that "It will have a positive effect on corporate identity" (4.41) and "Patient satisfaction will be increased" (4.38). As external independent audit firms will realize the mentioned assurance engagement, the reasons for managers to prefer such audits are definitely of significance.

Table 3.7 shows the findings related to 23 participants (44.2%), who avoid procuring assurance engagement service in conformity with the ISAE 3402. The most important reason for the participants to prevent purchasing the service is the "Audit being performed by the internal control unit of the hospital" (3.48), which is followed

Table 3.7 Findings on the reasons for managers to avoid assurance engagement service in conformity with the ISAE 3402

Cause	Total score	Average score	Effect rating
The audit is performed by the internal control unit of the hospital	80	3.48	**1**
The opinion that internal auditors can accomplish the same work	76	3.30	**2**
Inspection being costly	72	3.13	**3**
The audit being statutorily not obligatory	72	3.13	**3**
Avoidance of disclosure of information	71	3.09	4
The accreditation body performs the review	61	2.65	5
Negative impact on the profitability of the hospital	58	2.52	6
The ministry of health conducts the audit	51	2.22	7
The opinion that the ISAE 3402 is inapplicable	49	2.13	8

by the opinions that "the internal auditor can as well perform the same work" (3.30) and the audit "is statutorily not obligatory and is costly" (3.13).

Cross-examination method was employed to find out how many of the 34 participants, who reported their opinion that the ISAE 3402 standard is applicable, decided either to procure the assurance engagement or not; the findings obtained are shown in Table 3.8.

According to Table 3.8, 26 (76.47%) out of 34 participants who reported their opinion that the ISAE 3402 standard is applicable stated that they would procure the service, while 8 participants (23.53%) were observed to avoid purchasing the ceremony despite their statement that they think the ISAE 3402 is applicable. When the reasons for this situation were examined, the findings in Table 3.9 were obtained.

Table 3.9 shows the justifications submitted by the participants, who avoided procuring the service despite thinking that it is possible in terms of the industry for

Table 3.8 Comparison between the opinion of applicability and the demand

		Would you procure assurance engagement in the scope of the ISAE 3402?		
Do you think that the ISAE 3402 Standard is applicable?		Yes	No	Total
	Yes	26	**8**	34
	No	3	15	18
Total		29	23	52

Table 3.9 Findings on the justifications reported by the participants who avoided procuring the assurance engagement in conformity with the ISAE 3402 despite their opinion that the assurance engagement is applicable

	Average	Effect rating
The audit being statutorily not obligatory	**4.00**	**1**
Inspection being costly	**3.25**	**2**
The review is performed by the internal control unit of the hospital	**3.00**	**3**
The opinion that an internal auditor can accomplish the same work	2.63	4
Negative impact on the profitability of the hospital	2.62	5
Avoidance of disclosure of information	2.37	6
The accreditation body		
performs the audit 2,00 7 The ministry of health conducts the review	1.50	8
The opinion that the ISAE 3402 is inapplicable	1.25	9

the support services, outsourced by private hospitals in the scope of the ISAE 3402 standard, to be audited subject to assurance engagement by independent audit firms. Accordingly, Table 3.9 reveals that the primary factor affecting the participants' decisions on whether or not to procure the service is the fact that the assurance engagement in conformity with the ISAE 3402 standard is statutorily not obligatory. This factor is followed by the "Audit being costly" and "Audit being performed by the internal control unit of the hospital."

3.6.2 Conclusion and Suggestions

Healthcare organizations may choose to outsource clinical and nonclinical services by focusing on the main activities. For healthcare organizations and clients, it is of utmost importance to control and audit such services that are outsourced and referred to as support services. Therefore, the ministry of health implements various rules for the control and auditing of such outsourced services. Healthcare organizations carry out such monitoring and auditing activities by mean of units or persons they internally appoint. However, there is always doubt and distrust to some extent on such control and auditing activities conducted privately by healthcare organizations. The most significant reason for such suspicion is the fact that the healthcare organizations internally perform the inspections and audits. Whereas, outsourcing such control and auditing activities to an external entity or person will create a better environment of trust. Healthcare organizations may as well outsource the monitoring and auditing of the service they outsourced to a foreign audit firm. Independent audit firms perform their control and auditing activities by an audit standard setting out the methods for such controls and audits, issued by IFAC and put into effect by Public Company Accounting Oversight Board ("PCAOB") in Turkey. This standard was published with the code ISAE 3402 (TDS 3402 in Turkey).

In the scope of this study, the functioning mechanism of the audit process is described in the frame of the standard no 3402 for support services. Moreover, a survey was conducted by a private healthcare group to find out to what extent the managers of healthcare organizations are familiar with the standard no. 3402 and whether they are aware of the standard. Additionally, the rule was briefly introduced after that the participants were asked for their opinions about whether this standard can be implemented in the control and auditing of the support services.

As a result of the research, it was found out that all of the private healthcare organizations taking part in the study outsourced support services. The support services procured in hospitals were examined in two main groups as clinical and nonclinical services; accordingly, the most frequently outsourced services in the Private Healthcare Group participating in the research were seen to include "pathology, biochemistry lab., MR" in category of clinical services and "dining service, cleaning, cafeteria, security" in category of nonclinical services.

All of the participants taking part in the research reported that the support services outsourced are controlled and audited. It has been found out in the study that the control and auditing are performed by the hospital management (23.5%), quality department of the hospital (29.4%), internal control manager of the hospital (14.4%). The participants reporting that the independent audit firm performs the control and auditing are approximately at the rate of 1%. According to the results of the research, it is observed that the department responsible for the control and auditing activities varies depending on the hospital and there is no specific standard on this subject. It has been revealed that the control and auditing activities within healthcare organizations are most often realized in the form of periodic controls and quality audits. It is seen that powers and checks are performed intensively once a month. The methods used in controls and audits usually include observation, physical examination, and document inspection methods. The basis for controls and audits was seen to be "Checking of whether the organization complies with the terms of an agreement."

Around 55% of the participants (hospital managers) reported that they have never heard about the ISAE 3402 standard while 45% said that they have heard about the rule. However, it was revealed that only after being briefed on the ISAE 3402 did 65% of the participants state their opinion that this standard is applicable. The majority of the participants reported that they had not procured an audit service in a frame of the ISAE 3402, and 55% answered "yes" to the question "Would you procure such service?". The participants who think that the standard is applicable are seen to have not given the same response in respect of the decision to outsource such audit service to independent audit firms. Some parts of the participants stated that they would not outsource such service though thinking that the standard is applicable. The participants who gave a negative response to the question about procuring such service base their opinion on the fact that such types of services are internally controlled and audited in the hospital. Furthermore, the fact that the standard is statutorily not obligatory manifests itself as one of the major reasons. The participants who reported that they might procure assurance engagement service in conformity with the ISAE 3402 standard base their opinion on the statements that "the quality of service received will be increased," "it will have a positive effect on corporate identity," and "patient satisfaction will be increased."

As a result of the research, it was revealed that there is a need to further increase awareness of hospital managers on assurance engagement services in the frame of the ISAE 3402. It should be noted that assurance engagement services offered by independent audit firms in structure of the ISAE 3402 will improve the quality of services provided by healthcare organizations, such services to be procured will not bring a significant financial burden to the hospital but will increase client satisfaction and profitability, which consequently facilitate continuity of the healthcare organization in the long term.

To ensure users to be adequately informed, it is of significance that regulations of private healthcare organizations include descriptive information about audits regarding procured services rather than merely addressing details about the physical conditions for such service procurements. Turkish Ministry of Health should also regulate other outsourced services such as dining service, cleaning, cafeteria, security, laundry, pest control, etc., in similar quality to the restricted arrangement adopted in respect of auditing of information systems, which is, in fact, crucial for the development of healthcare as a quite significant sector. The assurance engagement service based on the ISAE 3402 will be of importance regarding raising awareness on the condition that it gains an obligatory discretionary status. Such a required state must be guaranteed through regulations to be issued by regulatory authorities or through promoting activities by associations or chambers operating in the healthcare industry rather than legal arrangements. This will merely be possible through the introduction of the ISAE 3402 standard and its advantages for healthcare organizations and through information meetings to be held on the subject.

References

Akbulut, Y., Tekeli, G., & Yildirim, T. (2012), Outsourcing in Turkish hospitals: A systematic review. *Ankara Sağlık Hizmetleri Dergisi, 11*(2), 25–33. http://doi.org/10.1501/Ashd.V11i2.1294.

Akyürek, Ç. E. (2013). *Sağlık Sisteminde Dış Kaynaklardan Yararlanma: T.C. Sağlık Bakanlığı Hastanelerinde Dış Kaynaklardan Yararlanma Uygulamalarının Değerlendirilmesi.* Unpublished Doctoral Dissertation, Institute of Social Sciences, Hacettepe University, Ankara.

Atabay, E. (2016). *Destek Hizmet Kuruluşlarındaki Kontrollere Yönelik Güvence Hizmetleri (GDS 3402): Türkiye'de Financial Sektördeki Farkındalığın Ölçümlenmesi.* Unpublished Doctoral dissertation, Institute of Social Sciences, Karadeniz Technical University, Trabzon.

Bailey, W., Wasson, R., & Raeside, R. (2002). Outsourcing in Edinburgh and the Lothians. *European Journal of Purchasing & Supply Management, 8*(2002), 83–95.

Bayindir, S. (2007). Yenilik Çalışmalarında Dış Kaynak Kullanımı. *Karaman Faculty of Economics and Administrative Sciences, Selçuk University Journal* (Special Edition on Local Economies), 241–250.

Beaumont, N., & Sohal, A. (2004). Outsourcing in Australia. *International Journal of Operations & Production Management, 24*(7), 688–700. https://doi.org/10.1108/01443570410541993.

Çakirer, M. A. (2003). *Sağlık Kurumları Bilgi Teknolojilerinde Dış Kaynaklardan (Outsourcing) Yararlanma Yöntemi Ve Hastane Yönetiminde Sağlayacağı Avantajlar*. ÜNAK'03: Changed Pathways for Information Retrieval and Second Symposium on Medical Information Management and Technologies, September 25. Retrieved October 24, 2017, from http://Kaynak.Unak.Org.Tr/Bildiri/Unak03/U03-24.Pdf.

Chen, Y., & Perry, J. (2004). *IT outsourcing: A primer for healthcare managers*. IBM Center For Healthcare Management Report. Retrieved October 24, 2017, from http://www.Floppybunny.Org/Robin/Web/Virtualclassroom/Chap12/S3/Articles/Healthcare_Outsourcing_Chen_Perry_2004.Pdf.

Court of Accounts, Republic of Turkey. (2014). Regularity Audit Guideline, Ankara.

Department of Auditing Services, Ministry of Health, Republic of Turkey. (2014). Private Hospitals Audit Guideline, Ankara.

Directive on the Procedures and Principles of Audits on Private Healthcare Institutions and Organizations by Provincial Directorates of Health, Authority's Approval Dated 05.08.2014 and Numbered 1449.

Döğücü, M. Ş., & Sayım, F. (2009). *Hastanelerde Dış Kaynak Kullanımı Kocaeli Örneği*. International Quality and Performance in Healthcare Congress, March 19–21, Antalya.

Foxx, W. K., Bunn, M. D., & Mccay, V. (2009). Outsourcing services in the healthcare sector. *Journal of Medical Marketing, 9*(1), 41–55.

Gözüküçük, M., & Çelik, Y. (2012). Sağlık Bakanlığı Hastanelerinde Dışarıdan Sağlık Hizmeti Alımı: Karşılaşılan Sorunlar Ve Çözüm Önerileri. *Hacettepe Journal of Health Administration, 15*(2), 1–25.

Guimarães, C. M., & Carvalho, J. C. (2011). Outsourcing in the healthcare sector—A state-of-the-art review. *Supply Chain Forum, 12*(1), 138–146.

Kalayci, Ş. (2014). *SPSS Uygulamalı Çok Değişkenli İstatistik Teknikleri* (6th ed.). Ankara: Asil Yayınları.

Kalkan, M., Sökmen, A., & Biyik, Y. (2015). Sağlık Hizmetlerinde Dış Kaynak Kullanımı: Ankara Halk Sağlığı Müdürlüğü Uygulaması. *Gazi Journal of Economics and Administrative Sciences, 17*(2), 36–56.

Karahan, A. (2009). Dış Kaynak Kullanımının Verimlilik Üzerine Etkisi (Hastane Yöneticileri Üzerine Bir Araştırma). *Balıkesir University Journal of the Institute of Social Sciences, 12*(21), 185–199.

Koszewska, M. (2004). Outsourcing as a modern management strategy. Prospects for its development in the protective clothing market. *AUTEX Research Journal, 4*(4), 228–231.

Ministry Of Health, Republic of Turkey. (2010). *Outsourcing practices in Turkish healthcare industry*. Ankara: School of Public Health.

Moschuris, S. J., & Kondylis, M. N. (2006). Outsourcing in public hospitals: A Greek perspective. *Journal of Health Organization and Management, 20*(1), 4–14.

Nazlioğlu, B., & Yar, C. E. (2016). Hastane İşletmelerinde Dış Kaynak Kullanımı – Maliyet İlişkisi: Literatür Taraması. *International Journal of Health Management and Strategies Research*, Cil:3, Sayı: 3, ISSN: 2149-6161, 72–80.

Onaydin Yildirim, M. (2011). *Dış Kaynak Kullanımı Ve Hastanelerde 3PL Lojistik*. Master's thesis, Bahçeşehir University, Institute of Science, Istanbul.

Sarivougioukas, J. C., & Vagelatos, A. Th. (2002). *IT outsourcing in the healthcare sector: The case of a state general hospital*. SIGCPR '02 Proceedings of the 2002 ACM SIGCPR Conference on Computer Personnel Research, Kristiansand, Norway, May 14–16, 58–61.

Sezer, A. (2009). *Hastane İşletmelerinde Dış Kaynak Kullanımının (Outsourcing) Hizmet Kalitesine Etkisi Ve Bir Uygulama*. Unpublished Master's thesis: Institute of Social Sciences, Dokuz Eylül University.

Türkeli, S. (2011). *Organizational quality measurement: War or/and peace of general and special to organization indicators*. 9th Health Management Congress Cyprus, 2011.

URL. Retrieved October 27, 2017, from http://www.Uhbd.Org/PDF/Sunumlar/JCI_ISO.Pdf.
URL. Retrieved October 27, 2017, from http://Ab.Org.Tr/Ab13/Sunum/216.Pdf.
Yalçin, İ., Akin, M., & Şeker, M. (2011). *Kamu Hastanelerinde Dış Kaynak Kullanımı: Kayseri Eğitim Ve Araştırma Hastanesi Örneği*. *KMÜ Social and Economic Research Journal, 13*(20), 83–92.
Yiğit, V. (2004). *Dış Kaynaklardan Yararlanma Ve Hastane İşletmelerinde Uygulanması*. Master's thesis, The Department of Business, The Institute of Social Sciences, Gazi University, Ankara.

Esra Atabay was born on June 17, 1981 in Kocaeli-Gölcük. He attended Çanakkale 18 March primary school, middle and high school Çanakkale M.A.E. Commerce and Anatolian Trade Vocational High School. In 1999, he earned the Business Administration Department of Anadolu University, Faculty of Economics and Administrative Sciences and completed her undergraduate studies in 2003. In the same year, he started her Master's degree in Department of Accounting at the same university. She completed her master's degree in 2007 by preparing the thesis entitled "Auditing in Public-Use Associations". She worked as a lecturer at Beykent University in 2007 and Kadir Has University in 2008. In 2009, she was appointed as a lecturer in Karadeniz Technical University, Vakfıkebir Vocational School. In 2011, she started her Ph.D. in Business Administration at the same university. In 2016, "Assurance auditing on controls at a service organization (ISAE 3402): Awareness Appraisal in the Financial Sector in Turkey", she completed her doctoral studies with a doctorate thesis. She was assistant professor in 2017. Currently, she is Head of Accounting and Tax Program with the title of Dr. Assistant Professor at Vocational School of Trabzon University. ATABAY is married and mother of two children. She speaks English.

Engin Dinç was born on September 25, 1966 in Trabzon-Arsin. In 1989, he was awarded the Business Administration Department of the Faculty of Economics and Administrative Sciences of Karadeniz Technical University. He completed his undergraduate studies in 1992 and started his graduate studies at the Institute of Social Sciences at the same university in the same year. In 1996, "Investigation of Computerized Accounting System and Computerized Evaluation of Accounting Practices in Turkey", he completed his master's degree with a thesis entitled. In the same year, he started his Ph.D. education in Business Administration in the same university and completed his education in 2001 with his doctoral dissertation entitled "Supervision of General Production Costs in Industrial Operations and an Application". He started his academic career at Karadeniz Technical University in 1993 as a research assistant. He received assistant professor in 2002, associate professor in 2010 and Professor Title in 2015. He still continues to serve as a faculty member of the Business Administration Department of the Faculty of Economics and Administrative Sciences of the Karadeniz Technical University. Speaks English.

Chapter 4
Revisiting Accountability: Corruption in Health Care in Developing Countries

Mohammad Nurunnabi

Abstract Corruption is epidemic in developing countries. Due to a lack of research, this chapter focuses on corruption in health care in developing countries. Based on an extensive literature search, this study finds that corruption is significantly increasing in health care. In addition, the local and international policymakers seriously address how to combat corruption in health care.

Keywords Corruption · Health care · Developing countries · Accountability

4.1 Introduction

> In too many countries, people are deprived of their most basic needs and go to bed hungry every night because of corruption, while the powerful and corrupt enjoy lavish lifestyles with impunity.—José Ugaz, Former Chair of Transparency International[1]

José Ugaz stresses that corruption is indeed like cancer. According to Transparency International (2016), "more than two-thirds of the 176 countries and territories were under the midpoint of the Corruption Perceptions Index 2016 scale of 0 (highly corrupt) to 100 (very clean)" (https://www.transparency.org). The global average score on an average is 43. This indicates the endemic corruption in a country's public sector. Table 4.1 presents the top ten clean countries from corruption. The Index also highlights that citizens face a tangible impact of corruption on a daily basis. Specifically, the higher ranked (very clean) countries tend to have the followings:

[1] https://www.transparency.org/news/feature/corruption_perceptions_index_2016.

M. Nurunnabi (✉)
Department of Accounting, College of Business Administration, Prince Sultan University, Rafha Street, PO Box – 66833, Riyadh 11586, Saudi Arabia
e-mail: mnurunnabi@psu.edu.sa; mohammad.nurunnabi@sant.ox.ac.uk

St Antony's College, University of Oxford, Oxford, UK

K. T. Çalıyurt (ed.), *Integrity, Transparency and Corruption in Healthcare & Research on Health, Volume I*, Accounting, Finance, Sustainability, Governance & Fraud: Theory and Application, https://doi.org/10.1007/978-981-15-1424-1_4

Table 4.1 Corruption perceptions index 2016 (High-ranked countries) [*Low level of corruption/very clean*]

2016 Rank	Country	2016 Score	2015 Score	2014 Score	2013 Score	2012 Score	Region
1	Denmark	90	91	92	91	90	Europe and Central Asia
1	New Zealand	90	91	91	91	90	Asia Pacific
3	Finland	89	90	89	89	90	Europe and Central Asia
4	Sweden	88	89	87	89	88	Europe and Central Asia
5	Switzerland	86	86	86	85	86	Europe and Central Asia
6	Norway	85	88	86	86	85	Europe and Central Asia
7	Singapore	84	85	84	86	87	Asia Pacific
8	Netherlands	83	84	83	83	84	Europe and Central Asia
9	Canada	82	83	81	81	84	Americas
10	Germany	81	81	79	78	79	Europe and Central Asia
10	Luxembourg	81	85	82	80	80	Europe and Central Asia
10	United Kingdom	81	81	78	76	74	Europe and Central Asia

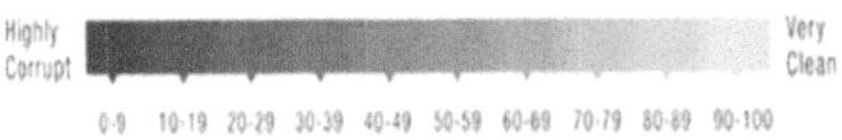

Source Transparency International (2016).
https://www.transparency.org/news/feature/corruption_perceptions_index_2016

- "higher degrees of press freedom,
- access to information about public expenditure,
- stronger standards of integrity for public officials, and
- independent judicial systems."

Source: https://www.transparency.org

On the other hand, the lower ranked countries (highly corrupt) in the index are plagued by ineffective public administration and judiciary. It is reported that the laws of anti-corruption are only in the books, but the implementation is often skirted or ignored. For instance, bribery and extortion, and misappropriation of funds are not uncommon in highly corrupt countries (Transparency International, 2016).

Corruption in health care in developing countries is also widespread. For instance, effective health services and accessibility remain black box in developing countries due to corruption (Habibov & Afandi, 2011; Habibov & Cheung, 2017; Habibov, Cheung, & Auchynnikava, 2017). However, the focus on corruption in health care is very limited in the literature.

The purpose of this chapter is to explore a review of corruption in health care in the developing countries. Accordingly, the chapter is organized as follows. The next Section briefly reviews the theory of corruption. Discussions on review of corruption in health care are explained in Sect. 3, followed by the conclusion.

4.2 Theory

Prior literature on corruption is enormous indeed. Various researchers detail the corruption from different angles. However, few papers demonstrate theory on corruption. Based on systematic literature, the following researchers demonstrate theoretical perspectives.

Country specific theory implications

- Bangladesh (Bhuiyan, 2011);
- Iraq (Chwastiak, 2013)
- Vietnam (Gueorguiev & Malesky, 2012)
- Former Soviet Bloc (Osipian, 2009) Former Soviet Bloc;
- Russia (Osipian, 2012) Russia
- Italy (Corrado & Rossetti, 2018; Sargiacomo, Ianni, D'Andreamatteo, & Servalli, 2015)
- Thailand (Ruengdet & Wongsurawat, 2015)
- Ukraine (Gorodnichenko & Peter, 2007).

Areas and themes of corruption

- e-governance; Poverty reduction; Public administration (Bhuiyan, 2011)
- Active corruption; Passive corruption; Bargaining power (Capasso & Santoro, 2018)
- Accountability; Critical; Public Interest; Fraud; Corruption; Auditing; Neoliberalism; Iraq (Chwastiak, 2013)
- Regions; Public sector; Panel data analysis (Corrado & Rossetti, 2018)
- Bribery; International business; Multinationals; Theory (Cuervo-Cazurra, 2016)
- Bribery; Public sector; Wage; Wage differentials; Consumption (Gorodnichenko & Peter, 2007)
- Vietnam; FDI; Unmatched count technique; Bribery; List question (Gueorguiev & Malesky, 2012)
- Higher education; Hierarchies (Osipian, 2009)
- Education; Economic growth; Reform; Transition (Osipian, 2012)
- Innovation; Quality certificates; Patents; Corruption; Trust; Firm; Heterogeneities (Paunov, 2016)
- Social model; Factors; Dependencies (Popova & Podolyakina, 2014)
- Economic rent; Rent-seeking; Agricultural price intervention (Ruengdet & Wongsurawat, 2015)

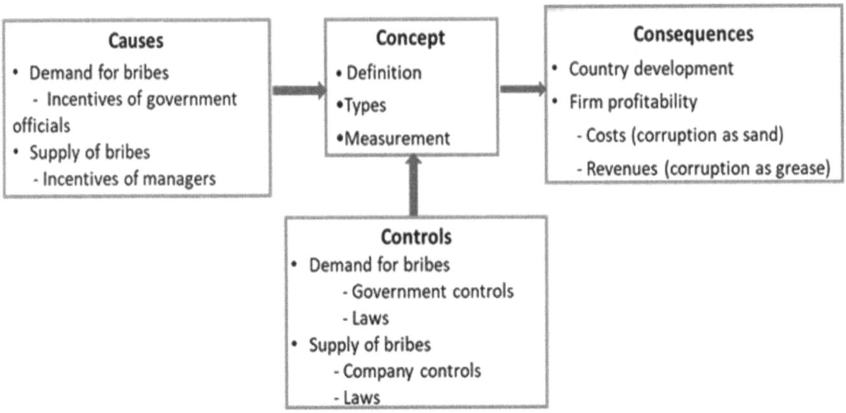

Fig. 4.1 Framework for analyzing corruption (Cuervo-Cazurra, 2016, p. 36)

- Information sharing; Extortionary corruption; Experiment; Crowdsourcing (Ryvkin, Serra, & Tremewan, 2017)
- Accounting; Government; Procurement (Sargiacomo et al., 2015).

In terms of the theory of corruption, Cuervo-Cazurra (2016) provided a framework for analyzing corruption in international business (see Fig. 4.1). He critically evaluates that

> corruption creates a laboratory for expanding international business studies because its illegal nature, the differences in perception about illegality, and the variation in the enforcement of laws against bribery across countries challenge (p. 35).

As presented in Fig. 4.1, several theories could be extended by analyzing the ethical power escape, neo-institutional theory, and resource-based view (e.g., illegal transaction costs minimization, corporate social irresponsibility, and illegal legitimacy). On the other hand, Capasso and Santoro (2018) explain two types of corruption: "the bargaining power is in the hands of the official (active corruption) or in the hands of the private agent (passive corruption)" (p. 104). Using Italian data, they reexamine the determinants of the aggregate level corruption. Discussing neoliberalism and accountability in US companies, Chwastiak (2013) finds the impact of the massive fraud as reported in the audit reports by the Office of the Special Inspector General for Iraq.

To explain the theories, some papers demonstrate the causes of corruption. For example, Corrado and Rossetti (2018) reveal that size and public spending at a local level may explain the level of corruption in Italy. Gorodnichenko and Peter (2007) report that public sector employees received 24–32% less wages than their private sector. Using the conditions of labor market equilibrium, they reveal that the extent of bribery is between US$460 million and 580 million in Ukraine (i.e., 0.9–1.2% of GDP of Ukraine in 2003).

Gueorguiev and Malesky (2012) examine the association between foreign direct investment (FDI) and corruption. They highlight that powerful external actors exert a corrupting influence in Vietnam. They suggest that economic openness may control the level of corruption in Vietnam. Popova and Podolyakina (2014) report that the corruption level is associated with the social model of the country.

Paunov (2016) examines the impacts of corruption in smaller and larger sized firms in 48 developing countries. They find that corruption likelihood is very low for those firms that obtained quality certificates. However, the adoption of quality certificates depends on the sustainable business environment. Ruengdet and Wongsurawat (2015) argue that risk management techniques and transparency are needed to control corruption in Thailand.

Corruption is also observed in higher education (Osipian, 2009, 2012). Osipian (2009) argues that the privatization and the decentralization in higher education in former Soviet Bloc result in the scope of increasing corruption. In a similar study, Osipian (2012) finds that the higher level of corruption in the education sector in Russia may harm economic productivity and lowering the level of human capital in Russia.

To explain the successful anti-corruption policies, Ryvkin et al. (2017) discuss Citizens' behavior and Officials' behavior in India. Using the voluntary and anonymous reports of bribe demands (I paid a bribe website—http://ipaidabribe.com/, this website was first launched in India in 2010), suggest a more effective platform may reduce corruption. They also reveal that citizens hardly post false information in "I paid a bribe website."

Further, based on a longitudinal analysis of the 22-year period in Italy, Sargiacomo et al. (2015) critically evaluated accounting-based anti-corruption model to control corrupt behaviors. They find a vast network of corrupt businessmen, politicians, and bureaucrats implicated in the expose of fraud case, "Clean Hands" ("Mani Pulite").

Several Researchers put forward to tackle corruption in health care. For instance, Bhuiyan (2011) examines the role of e-governance in controlling corruption and reducing poverty. He stresses that government employees (public administration) may play an instrumental role in achieving success.

4.3 Discussion

According to Transparency International (2016), corruption varies from region to region (see Figs. 4.2, 4.3, 4.4, 4.5 and 4.6): For example,

- **Americas**: The region is fighting against corruption. Haiti and Guatemala are still highly corrupt countries.
- **Asia Pacific**: Most of the countries in this region are at the bottom half of the Index. This is mainly because of the lack of accountability of governments, lack of oversight, inactive civil society, and lack of anti-corruption action. Afghanistan and Bangladesh were at the bottom category.
- **Europe and Central Asia**: The region seems that there were no drastic changes in Corruption Perceptions Index comparing to the prior year, with only a few

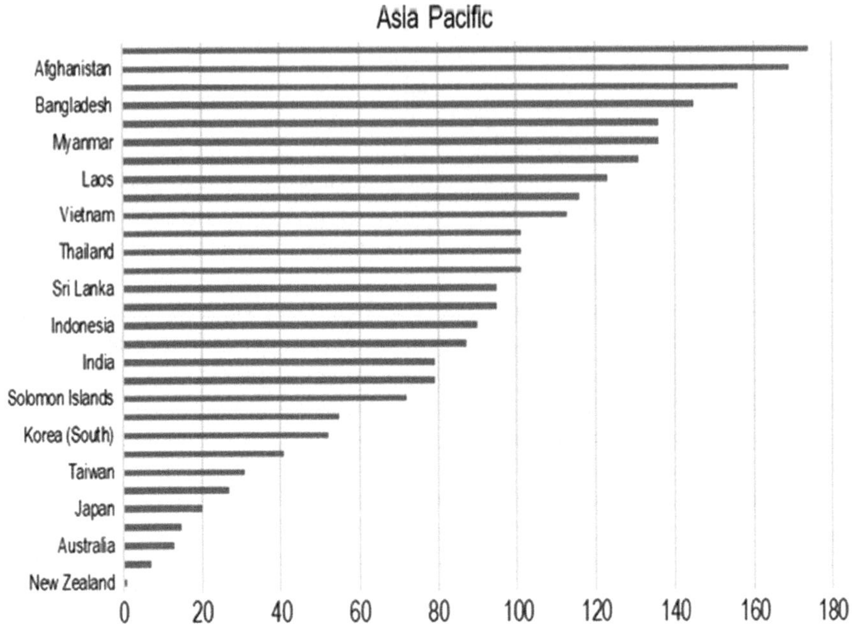

Fig. 4.2 Corruption ranking of Asia Pacific in 2016

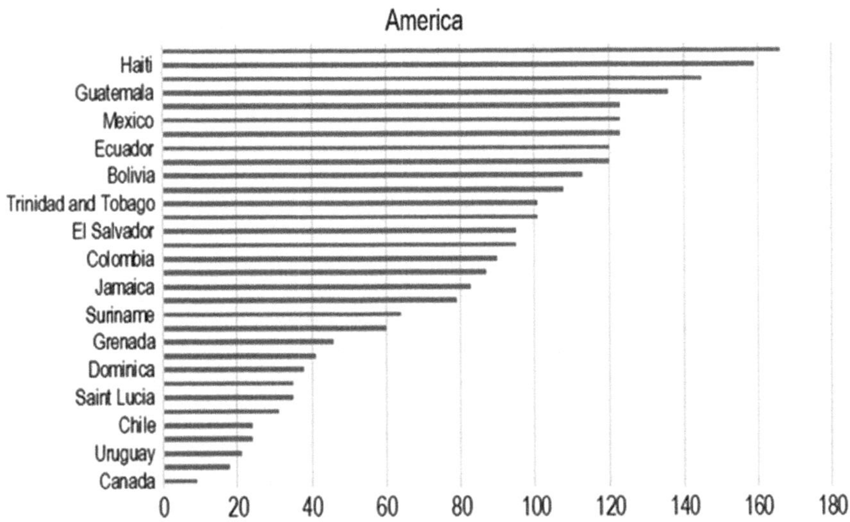

Fig. 4.3 Corruption ranking of Americas (North and South America) in 2016

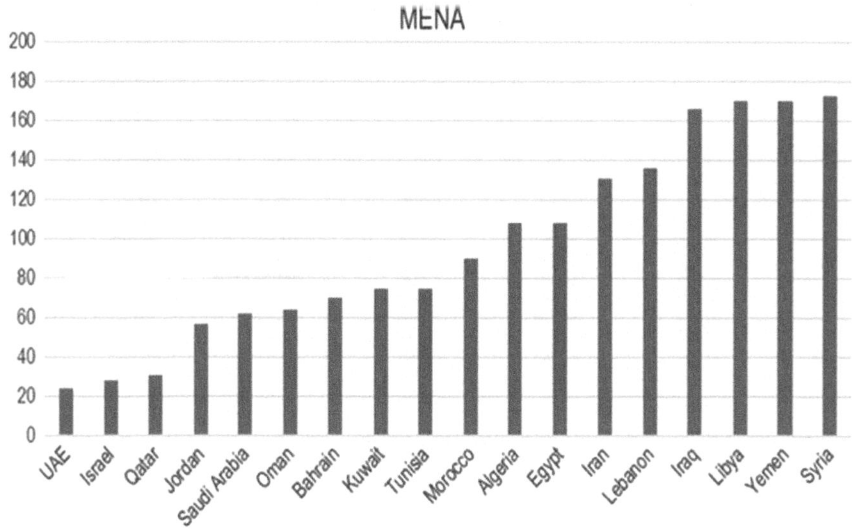

Fig. 4.4 Corruption ranking of MENA (Middle East and North Africa) in 2016

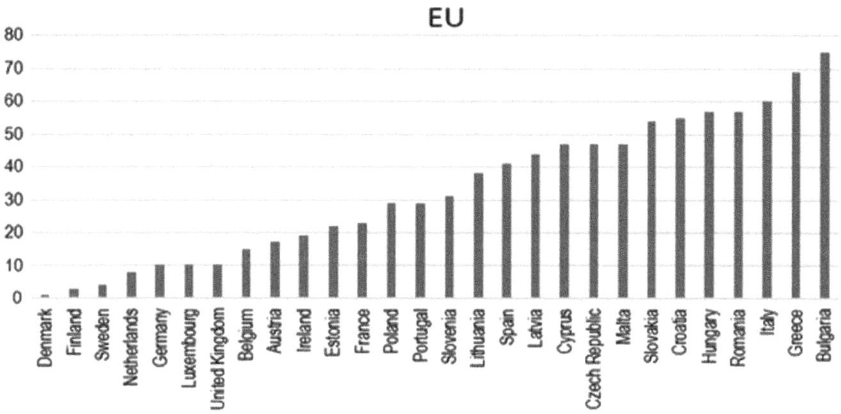

Fig. 4.5 Corruption ranking of EU (European Union) in 2016

exceptions. However, Transparency International (2016) reports that "this does not mean that the region is immune from corruption."

- **Middle East and North Africa**: There were no significant progress to fight corruption in the Arab region. Most of the Arab countries (90%) have scored below 50, which is a worsening grade (particularly, Libya, Yamen, Syria, Iraq, and Lebanon are at the lowest rank).

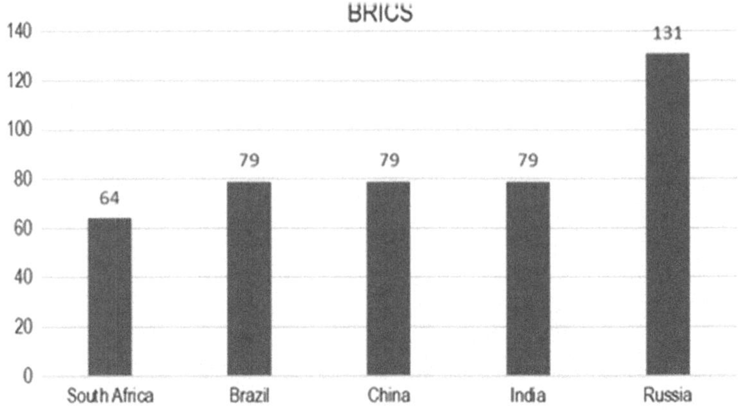

Fig. 4.6 Corruption ranking of BRICS in 2017 (Brazil, Russia, India, China, and South Africa)

- **Sub-Saharan Africa**: The region has seen an improvement in the Index. For instance, in Ghana, voters voiced their dissatisfaction with the government's corruption record at the polls (i.e., the first time in Ghana's history, an incumbent president was voted out).

Based on an extensive literature search, there were 31 articles were found from the Elsevier database directly related to corruption and health care. Table 4.2 shows the list of authors and Table 4.2 provides the major findings of the articles.

4.4 Conclusions

> The country's [India] doctors and medical institutions live in an 'unvirtuous circle' of referral and kickback that poisons their integrity and destroys any chance of a trusting relationship with their patients. Given these practices, it is no surprise that investigations and procedures are abused as a means of milking patients (Berger, 2014, g3169).

In this chapter, I reviewed corruption in health care in developing countries, going beyond the traditional review of the literature. Undoubtedly, corrupt practices represent a growing social malaise worldwide. This problem is in developing countries are pervasive. Healthcare access and corruption remain key challenges in health care (Winchester & King, 2018). For instance, Andaleeb (2001) reports that one may be hard-pressed to find a service sector in which there is no corruption and health care is also not immune from it.

Informal payments, bribes, extra payments for the access healthcare facilities are widely reported in the literature. Based on the review, I suggest that the use of information technology could reduce soaring healthcare costs and enhancing service quality (Turan & Palvia, 2014). It was found that falsifying medical reports were not uncommon practices in many developing countries. The effective use of technology

Table 4.2 Major findings of Articles on corruption and health care

Author(s)	Major findings
Andaleeb (2001)	The study argues that patients' voice plays a greater role in healthcare service in developing countries. The study identified service quality factors in the context of Bangladesh. Based on a field survey, several dimensions were reported including responsiveness, assurance, communication, discipline, and baksheesh. The study also highlights the corruption issues. For instance, healthcare officials obtained money from patients for the mere facilitation of due services
Cavalieri, Guccio, and Rizzo (2017)	The study utilized two-stage Data Envelopment Analysis (DEA) and finds that the performance of Italian public contracts for healthcare infrastructures is significantly affected by 'environmental' corruption. The findings suggest having 'qualified' public contracting authorities in Italy
Chereches, Ungureanu, Sandu, and Rus (2013)	Discussing the informal payments in health care, the study reviews the definitions based on reviewing prior research and propose the definition. The study followed a search strategy to identify the relevant papers related to informal payments from various sources (PubMed, ScienceDirect, Econlit, Google Scholar, and EconPapers). The study finds that the term "informal payments" has been used as "under-the-counter payments", "under-the-table payments", and "envelope payments". However, there was a lack of a unified definition. The study stressed that informal payments are illegal
Chuang, Sung, Chao, Bai, and Chang (2013)	Based on the longitudinal dataset and regression models, the study addressed some challenges in sub-Saharan Africa including a lack of drugs and supplies, a lack of enforcement of regulation, high customs duties and taxes, and corruption
Dodd, King, Humphries, Little, and Dewey (2016)	In identifying the priority areas for government health policies and programs, the study examines the self-reported healthcare preferences and experiences, health literacy, and morbidity in Tamil Nadu, India. The study was conducted on 300 household surveys and 66 semi-structured interviews in 26 rural villages in Krishnagiri district (Tamil Nadu) The study finds that government facilities were mistrusted. The major challenges to healthcare access are high treatment cost and corruption
Egharevba and Atkinson (2016)	The study evaluates the ethical issues in relation clinical trials (pharmaceutical industry sponsored) in sub-Saharan Africa. Based on the survey questionnaire and interviews, they find that corruption was a significant issue
Global Burden of Disease Health Financing Collaborator Network (2018)	The report discusses the decentralization reforms in low- and middle-income countries with reference to health politics in Indonesian local government. The report highlights that "focus on democratic accountability and good governance may be insufficient to explain major policy outcomes associated with decentralization…..to understand policy trajectories over a longer time frame, relations between politicians at different levels of government become the crucial factor" (p. 1783) Based on qualitative and quantitative data from nearly 400 Indonesian districts and provinces, the report suggests that policy cooperation is needed across levels of government

(continued)

Table 4.2 (continued)

Author(s)	Major findings
Houngbo, Zweekhorst, Bunders, Coleman, Medenou, Dakpanon, and Buning (2017)	Based on two surveys from 2008 to 2010 in Benin public health sector (377 questionnaires and 259 interviews), the study examines the root causes and solutions of Healthcare Technology Management. They reveal that the major problems are high- and low-level of corruption, and unwillingness of the policymakers
Hu and Mossialos (2016)	The study reviews the pharmaceutical spending in China and the quality, pricing, and affordability of drugs in China. They mention that "China's current pharmaceutical policies interact in such a way to create dysfunction in the form of high prices, low drug quality, irrational prescribing and problems with access" (p. 519). They highlight several challenges including regulatory fragmentation, a lack of transparency, and corruption
Liang and Mirelman (2014)	The study investigates a complex relationship across government health expenditure (GHE) of 120 countries from 1995 to 2010. Interestingly, they find that "Corruption is associated with less GHE in developing countries, but with higher GHE in developed countries. …that development assistance for health (DAH) is fungible with domestically financed government health expenditure (DGHE)" (p. 161)
Rönnerstrand and Lapuente (2017)	Investigating corruption and antibiotic use at the subnational level (using data from the European Quality of Government Index and consumption of antibiotics from 2009 Special Euro Barometer), the study uses measures of corruption ("prevalence of corruption in the health sector and prevalence of bribes in the society"). The study reveals that there is a strong positive association between both measures of corruption and antibiotics use in Europe
Singh and Wang (2013)	The study evaluates the clinical trials and Good Clinical Practice (GCP). Based on the factors including the country's GDP, patent applications, healthcare expenditure, healthcare infrastructure, corruption, innovation, the study reveals that there is a significant correlation with the number of clinical trials in emerging markets
Ssozi and Amlani (2015)	The study examines the effectiveness of health expenditure from 1995 to 2011 in 43 nations of Sub-Saharan Africa using the General Method of Moments (GMM) technique. The study finds that public service delivery improvement could result in effective health expenditure. However, they stress that low government effectiveness, and corruption may hinder the government's effective health expenditure
Stepurko, Pavlova, Gryga, Murauskiene, and Groot (2015)	The study examines the patterns of gifts, tips, and bribes paid by patients for healthcare services in post-Soviet and post-communist countries (e.g., Lithuania, Ukraine, and Poland). They find that "a lower share of informal patient payments as well as a prevalence of more negative attitudes towards informal patient payments in Poland compared to Lithuania and Ukraine" (p. 46). Comparing inpatient and outpatient healthcare services, it is found that informal payments ("bribes and fees") are common to in-patient healthcare services

(continued)

Table 4.2 (continued)

Author(s)	Major findings
Turan and Palvia (2014)	In the experience of healthcare facilities in Turkey, the study collected surveys of senior hospital managers. They argue that IT could be helpful in enhancing service quality and reducing healthcare costs. Several issues were reported including privacy, quality, security, and the implementation of electronic medical records
Turcotte-Tremblay, Gali-Gali, Allegri, and Ridde (2017)	Evaluating the performance-based financing (PBF) in Africa, the study collected 92 interviews and non-participant observation from January to May 2016. Using diffusion of innovations theory, the study finds that there was evidence of a financial incentive to falsify reports and corruption
Winchester and King (2018)	The study acknowledged that South Africa has followed international strategies of healthcare decentralization. The study collected data from focus group, survey, interviews, and clinic data in the northeast region of South Africa for the period 2013–2016. They find that the decentralization strategy fails to resolve the problems of healthcare access in the current period due to corruption and mismanagement

could stop such practices. In addition, Patients' perceptions about healthcare facilities should be heard regularly on various platforms through research and social media. Unfortunately, this issue has been paid little attention to healthcare providers in developing countries. Suitable regulations, implementation of such regulations and policy, accountability framework of healthcare providers coupled with (patients' voice may decrease the level of corruption in developing countries.

Future studies can take the suggestions provided in this chapter and analyze them in more detail, helping extend not only our understanding of corruption in health care, but also using theory and policy prescriptions based on the contextual factors.

References

Andaleeb, S. S. (2001). Service quality perceptions and patient satisfaction: A study of hospitals in a developing country. *Social Science and Medicine, 52*(9), 1359–1370. https://doi.org/10.1016/S0277-9536(00)00235-5.

Berger, D. (2014). Corruption ruins the doctor-patient relationship in India. *BMJ, 348*, g3169. https://doi.org/10.1136/bmj.g3169.

Bhuiyan, S. H. (2011). Modernizing Bangladesh public administration through e-governance: Benefits and challenges. *Government Information Quarterly, 28*(1), 54–65. https://doi.org/10.1016/j.giq.2010.04.006.

Capasso, S., & Santoro, L. (2018). Active and passive corruption: Theory and evidence. *European Journal of Political Economy, 52*, 103–119. https://doi.org/10.1016/j.ejpoleco.2017.05.004.

Cavalieri, M., Guccio, C., & Rizzo, I. (2017). On the role of environmental corruption in healthcare infrastructures: An empirical assessment for Italy using DEA with truncated regression approach. *Health Policy, 121*(5), 515–524. https://doi.org/10.1016/j.healthpol.2017.02.011.

Chereches, R. M., Ungureanu, M. I., Sandu, P., & Rus, I. A. (2013). Defining informal payments in healthcare: A systematic review. *Health Policy, 110*(2–3), 105–114. https://doi.org/10.1016/j.healthpol.2013.01.010.

Chuang, Y.-C., Sung, P.-W., Chao, H. J., Bai, C.-H., & Chang, C.-J. (2013). A longitudinal ecological study of the influences of political, economic, and health services characteristics on under-five mortality in less-developed countries. *Health & Place, 23,* 111–121. https://doi.org/10.1016/j.healthplace.2013.05.007.

Chwastiak, M. (2013). Profiting from destruction: The Iraq reconstruction, auditing and the management of fraud. *Critical Perspectives on Accounting, 24*(1), 32–43. https://doi.org/10.1016/j.cpa.2011.11.009.

Cockerham, W. C., Hamby, B. W., Hankivsky, O., Baker, E. H., & Rouhani, S. (2017). Self-rated health and barriers to healthcare in Ukraine: The pivotal role of gender and its intersections. *Communist and Post-Communist Studies, 50*(1), 53–63. https://doi.org/10.1016/j.postcomstud.2017.01.001.

Corrado, G., & Rossetti, F. (2018, in press). Public corruption: A study across regions in Italy. *Journal of Policy Modeling*. https://doi.org/10.1016/j.jpolmod.2018.01.001.

Cuervo-Cazurra, A. (2016). Corruption in international business. *Journal of World Business, 51*(1), 35–49. https://doi.org/10.1016/j.jwb.2015.08.015.

Dodd, W., King, N., Humphries, S., Little, M., & Dewey, C. (2016). Self-reported morbidity and health service utilization in rural Tamil Nadu, India. *Social Science and Medicine, 161,* 118–125. https://doi.org/10.1016/j.socscimed.2016.05.035.

Egharevba, E., & Atkinson, J. (2016). The role of corruption and unethical behaviour in precluding the placement of industry sponsored clinical trials in sub-Saharan Africa: Stakeholder views. *Contemporary Clinical Trials Communications, 3,* 102–110. https://doi.org/10.1016/j.conctc.2016.04.009.

Ekbrand, H., & Halleröd, B. (2018). The more gender equity, the less child poverty? A multilevel analysis of malnutrition and health deprivation in 49 low- and middle-income countries. *World Development, 108,* 221–230. https://doi.org/10.1016/j.worlddev.2018.01.028.

Footman, K., Roberts, B., Mills, A., Richardson, E., & McKee, M. (2013). Public satisfaction as a measure of health system performance: A study of nine countries in the former Soviet Union. *Health Policy, 112*(1–2), 62–69. https://doi.org/10.1016/j.healthpol.2013.03.004.

Fossati, F. (2016). Beyond "Good Governance": The Multi-level Politics of Health Insurance for the Poor in Indonesia. *World Development, 87,* 291–306. https://doi.org/10.1016/j.worlddev.2016.06.020.

Giménez, V., Prieto, W., Prior, D., & Tortosa-Ausina, E. (2018, in press). Evaluation of efficiency in Colombian hospitals: An analysis for the post-reform period. *Socio-Economic Planning Sciences*. http://www.sciencedirect.com/science/article/pii/S0038012117301714.

Global Burden of Disease Health Financing Collaborator Network. (2018). Trends in future health financing and coverage: Future health spending and universal health coverage in 188 countries, 2016–40. *The Lancet, 391*(10132), 1783–1798. https://doi.org/10.1016/S0140-6736(18)30697-4.

Gorodnichenko, Y., & Peter, K. S. (2007). Public sector pay and corruption: Measuring bribery from micro data. *Journal of Public Economics, 91*(5–6), 963–991. https://doi.org/10.1016/j.jpubeco.2006.12.003.

Gueorguiev, D., & Malesky, E. (2012). Foreign investment and bribery: A firm-level analysis of corruption in Vietnam. *Journal of Asian Economics, 23*(2), 111–129. https://doi.org/10.1016/j.asieco.2011.11.006.

Habibov, N. (2016). Effect of corruption on healthcare satisfaction in post-soviet nations: A cross-country instrumental variable analysis of twelve countries. *Social Science and Medicine, 152,* 119–124. https://doi.org/10.1016/j.socscimed.2016.01.044.

Habibov, N., & Cheung, A. (2017). Revisiting informal payments in 29 transitional countries: The scale and socio-economic correlates. *Social Science and Medicine, 178,* 28–37. https://doi.org/10.1016/j.socscimed.2017.02.003.

Habibov, N. N., & Afandi, E. N. (2011). Self-rated health and social capital in transitional countries: Multilevel analysis of comparative surveys in Armenia, Azerbaijan, and Georgia. *Social Science and Medicine, 72*(7), 1193–1204. https://doi.org/10.1016/j.socscimed.2011.01.035.

Habibov, N., Cheung, A., & Auchynnikava, A. (2017). Does social trust increase willingness to pay taxes to improve public healthcare? Cross-sectional cross-country instrumental variable analysis. *Social Science and Medicine, 189,* 25–34. https://doi.org/10.1016/j.socscimed.2017.07.023.

Houngbo, P. T., Zweekhorst, M., Bunders, J., Coleman, H. L. S., Medenou, D., Dakpanon, L., et al. (2017). The root causes of ineffective and inefficient healthcare technology management in Benin public health sector. *Health Policy and Technology, 6*(4), 446–456. https://doi.org/10.1016/j.hlpt.2017.06.004.

Hu, J., & Mossialos, E. (2016). Pharmaceutical pricing and reimbursement in China: When the whole is less than the sum of its parts. *Health Policy, 120*(5), 519–534. https://doi.org/10.1016/j.healthpol.2016.03.014.

Iacobuţă, A.-O. (2012). Patients' evaluation of access and quality of healthcare in EU countries. A comparative analysis. *Procedia—Social and Behavioral Sciences, 62,* 896–900. https://doi.org/10.1016/j.sbspro.2012.09.151.

Kostyak, L., Shaw, D. M., Elger, B., & Annaheim, B. (2017). A means of improving public health in low- and middle-income countries? Benefits and challenges of international public–private partnerships. *Public Health, 149,* 120–129. https://doi.org/10.1016/j.puhe.2017.03.005.

Liang, L.-L., & Mirelman, A. J. (2014). Why do some countries spend more for health? An assessment of sociopolitical determinants and international aid for government health expenditures. *Social Science and Medicine, 114,* 161–168. https://doi.org/10.1016/j.socscimed.2014.05.044.

Medhekar, A. (2014). Public-private partnerships for inclusive development: Role of private corporate sector in provision of healthcare services. *Procedia—Social and Behavioral Sciences, 157,* 33–44. https://doi.org/10.1016/j.sbspro.2014.11.007.

Osipian, A. L. (2009). Corruption hierarchies in higher education in the former Soviet Bloc. *International Journal of Educational Development, 29*(3), 321–330. https://doi.org/10.1016/j.ijedudev.2008.08.006.

Osipian, A. L. (2012). Education corruption, reform, and growth: Case of Post-Soviet Russia. *Journal of Eurasian Studies, 3*(1), 20–29. https://doi.org/10.1016/j.euras.2011.10.003.

Paunov, C. (2016). Corruption's asymmetric impacts on firm innovation. *Journal of Development Economics, 118,* 216–231. https://doi.org/10.1016/j.jdeveco.2015.07.006.

Popova, Y., & Podolyakina, N. (2014). Pervasive impact of corruption on social system and economic growth. *Procedia—Social and Behavioral Sciences, 110,* 727–737. https://doi.org/10.1016/j.sbspro.2013.12.917.

Rodriguez, C. A. (2016). Challenges to effectiveness in public health organizations: The case of the Costa Rican Health Ministry. *Journal of Business Research, 69*(9), 3859–3868. https://doi.org/10.1016/j.jbusres.2015.11.018.

Rönnerstrand, B., & Lapuente, V. (2017). Corruption and use of antibiotics in regions of Europe. *Health Policy, 121*(3), 250–256. https://doi.org/10.1016/j.healthpol.2016.12.010.

Rosenberg, D., Kozlov, V., & Libman, A. (2018). Political regimes, income and health: Evidence from sub-national comparative method. *Social Science Research, 72,* 20–37. https://doi.org/10.1016/j.ssresearch.2018.02.001.

Ruengdet, K., & Wongsurawat, W. (2015). The mechanisms of corruption in agricultural price intervention projects: Case studies from Thailand. *The Social Science Journal, 52*(1), 22–33. https://doi.org/10.1016/j.soscij.2014.12.011.

Ryvkin, D., Serra, D., & Tremewan, J. (2017). I paid a bribe: An experiment on information sharing and extortionary corruption. *European Economic Review, 94,* 1–22. https://doi.org/10.1016/j.euroecorev.2017.02.003.

Sargiacomo, M., Ianni, L., D'Andreamatteo, A., & Servalli, S. (2015). Accounting and the fight against corruption in Italian government procurement: A longitudinal critical analysis (1992–2014). *Critical Perspectives on Accounting, 28,* 89–96. https://doi.org/10.1016/j.cpa.2015.01.006.

Schneider, S. M., & Popic, T. (2018). Cognitive determinants of healthcare evaluations—A comparison of Eastern and Western European countries. *Health Policy, 122*(3), 269–278. https://doi.org/10.1016/j.healthpol.2017.12.012.
Singh, R., & Wang, O. (2013). Clinical trials in "emerging markets": Regulatory considerations and other factors. *Contemporary Clinical Trials, 36*(2), 711–718. https://doi.org/10.1016/j.cct.2013.09.006.
Ssozi, J., & Amlani, S. (2015). The effectiveness of health expenditure on the proximate and ultimate goals of healthcare in sub-Saharan Africa. *World Development, 76,* 165–179. https://doi.org/10.1016/j.worlddev.2015.07.010.
Stepurko, T., Pavlova, M., Gryga, I., Murauskiene, L., & Groot, W. (2015). Informal payments for health care services: The case of Lithuania, Poland and Ukraine. *Journal of Eurasian Studies, 6*(1), 46–58. https://doi.org/10.1016/j.euras.2014.11.002.
Transparency International. (2016). *Corruption perceptions index 2016*. Retrieved March 15, 2017, from https://www.transparency.org/news/feature/corruption_perceptions_index_2016.
Turan, A. H., & Palvia, P. C. (2014). Critical information technology issues in Turkish healthcare. *Information & Management, 51*(1), 57–68. https://doi.org/10.1016/j.im.2013.09.007.
Turcotte-Tremblay, A. M., Gali-Gali, I. A., Allegri, M. D., & Ridde, V. (2017). The unintended consequences of community verifications for performance-based financing in Burkina Faso. *Social Science and Medicine, 191,* 226–236. https://doi.org/10.1016/j.socscimed.2017.09.007.
Winchester, M. S., & King, B. (2018). Decentralization, healthcare access, and inequality in Mpumalanga, South Africa. *Health & Place, 51,* 200–207. https://doi.org/10.1016/j.healthplace.2018.02.009.

Mohammad Nurunnabi Ph.D., CMA, SFHEA FRSA FAIA (Acad), is the Aide to the Rector on Research and Internationalization, Chair of the Department of Accounting, and an Associate Professor in Accounting at Prince Sultan University, Saudi Arabia. He is an Academic Visitor of St Antony's College, University of Oxford, UK. Previously he has taught in University of East London, United Kingdom (UK), University of Bedfordshire (UK), and Edge Hill University (UK). Prior to his academic career, he worked as a financial team member in ADVFN Plc (covers in excess of 70 stock exchanges from across the globe), London, UK.

Part II
National & International Regulation on Fraud Prevention in Healthcare & Pharmaceutical

Chapter 5
Corruption of the Canadian Drug Regulatory System

Joel Lexchin

Abstract The thesis of this chapter is that the long-standing relationship between Health Canada and the pharmaceutical industry has corrupted the Canadian regulatory system. Health Canada is much more geared to the philosophy that drug regulation is primarily a commercial activity to help manufacturers get their products to the market as quickly as possible than to the idea that regulation should protect public health by ensuring that medicines that reach the market are efficacious and sufficiently safe to ensure that the benefit–harm balance is favourable. Corruption is a "loaded" term. In this case, it does not mean that bribes have been paid to Health Canada employees or that anything illegal has taken place, rather that the regulatory system has been corrupted from its primary purpose of serving the public interest to serving commercial interests. This thesis will be explored first by reviewing the history of industry–Health Canada relations and how it demonstrates an ongoing cooperative relationship whereby Health Canada has been willing to delegate regulatory responsibilities to the industry and prioritize industry's views over those of others. The chapter then looks at how clinical trials are regulated and how the desire of industry for trials to be done quickly and as inexpensively as possible means that economic values can outweigh scientific ones. The chapter then turns to the regulatory review process and how the quality of evidence that Health Canada accepts and how quickly it reviews that information impacts on the efficacy and safety of drugs that reach the market. Promotion has a significant effect on how drugs are prescribed by physicians and the chapter critically examines how Health Canada has turned over the regulation to either industry or bodies that are closely aligned with industry. Finally, the chapter looks at the deficiencies in how safety is monitored once drugs are being prescribed, sold and used by patients.

Keywords Corruption · Canadian drug regulatory system

J. Lexchin (✉)
Faculty of Health, York University, Toronto, Canada
e-mail: jlexchin@yorku.ca

K. T. Çalıyurt (ed.), *Integrity, Transparency and Corruption in Healthcare & Research on Health, Volume I*, Accounting, Finance, Sustainability, Governance & Fraud: Theory and Application, https://doi.org/10.1007/978-981-15-1424-1_5

5.1 Introduction

At the extreme, there are two views of the function of a drug regulatory system. One is that its primary job is to protect the public and ensure that the medicines that reach the market are efficacious and sufficiently safe to ensure that the benefit–harm balance is favourable. The second view is that drug regulation is primarily a commercial activity to help manufacturers get their products to the market as quickly as possible. The first view aligns with the precautionary principle, i.e., when there is doubt about the safety of a product, even in the absence of definitive scientific data, regulators should proceed cautiously and either hold off on approving a new drug or else only allow it onto the market under restricted circumstances. The second view aligns with a risk management approach to regulation, i.e., if risk has not been shown then the product should be approved and only after risk(s) have been identified should action be taken.

The thesis of this chapter is that the long-standing relationship between Health Canada and the pharmaceutical industry has corrupted the Canadian regulatory system such that it is much more aligned to the second view than the first. Corruption is a "loaded" term. In this case, it does not mean that bribes have been paid to Health Canada employees or that anything illegal has taken place, rather that the regulatory system has been corrupted from its primary purpose of serving the public interest to serving commercial interests. This thesis will be explored first by reviewing the history of industry–Health Canada relations and then by looking at how clinical trials are regulated, the regulatory review process and its implications for the efficacy and safety of drugs that reach the market, how those drugs are promoted and finally how safety is monitored once drugs are being prescribed, sold and used by patients.

5.2 Folie à Deux: Health Canada and the Pharmaceutical Industry

The relationship between the Canadian government and the brand-name multinational pharmaceutical industry, as represented by its national organization, Innovative Medicines Canada, has developed through a path, termed clientele pluralism. As Atkinson and Coleman described it (Atkinson & Coleman, 1985, 1989), in such a system the state concentrates power in a single agency, in the case of drug regulation that agency is Health Canada, that either does not possess the ability, or lacks interest in developing the capacity, to ensure safe and effective medications on its own. From the very start of drug regulation in Canada, the predecessors to Health Canada had limited staffing and had to rely heavily on the industry for self-regulation (Potter, 1966). Therefore, some authority must be voluntarily relinquished to the drug manufacturers to pursue objectives with which officials are in broad agreement. This low degree of autonomy on the part of Health Canada may partly exist because of a lack of expertise in the agency to address drug regulation, but primarily it is due to the political orientation of the state. In other words, the state sees many of its interests as synonymous with those of the industries that it is charged to regulate.

Once Canada embarked on a regulatory system defined by clientele pluralism, it seldom strayed from this approach. Speaking to the House of Commons in 1940, the Minister of Health Ian McKenzie explained what he considered to be the general approach to the administration of the Food and Drugs Act (F&DAct). "The statute… is not entirely punitive. Much of the control is exercised by preventing action in which the industry and trade co-operate." Six years later, another Minister of Health explained that "we try to secure enforcement through co-operation, first with the representatives of the various trade associations and, second with the people in business." Again in 1953, a third minister commented "I cannot speak too highly of the co-operation we have had from industry generally in the administration of this [Food and Drugs] Act" (Potter, 1966) (p. 9). The underlying approach is perhaps best summed up in the 1975 quote from Dr. A. B. Morrison, the Assistant Deputy Minister of the Health Protection Branch (the branch that at the time was in charge of drug regulation), of the Department of National Health and Welfare. "We prefer to work co-operatively with responsible manufacturers and to encourage voluntary compliance by industry. We try to avoid unnecessary confrontation and adversary proceedings insofar as possible" (Morrision, 1975) (p. 642).

The post-thalidomide 1962 revisions to the F&DAct required companies to actually prove that their drugs were effective before they could be approved for marketing. In introducing these amendments, the Minister of Health was careful to emphasize that they were not intended "to affect the right of a manufacturer to inform a physician of a new product or to deny him the right to make a new product available by way of sample. This would be an unwarranted interference with the professions and with the industry" (House of Commons Debates: Official Report, 1962) (p. 977).

As Health Minister, Judy LaMarsh, delivered the welcoming address at the 1964 5th Annual General Meeting of the predecessor to Innovative Medicines Canada and warmly applauded the government–industry relationship. During her speech, she noted that the "task [of the Director of the Food and Drugs Directorate] would be immeasurably more difficult if he did not have access to the combined knowledge of the industry and receive its support." She went on to further say "the role of a responsible trade association, in my view, is the advice and assistance it can offer to the government in carrying out its responsibility to the Canadian people…In the past [we have received from you] valuable help and assistance in the development and administration of our drug regulations…In the formulation of our present Act, committees of your Association met with officers of the Department and worked out matters which are now reflected in the provisions of the law itself" (Anderson, 1977) (p. 298). As Anderson notes (Anderson, 1977), the speech clearly indicates that the industry and bureaucrats in the Department were well known to each other and reveals the degree of influence that the industry had on policymaking.

A series of articles on prescription drug regulation in the Montreal Gazette in 1982 showed that the attitude Health Canada officials brought to their approach to drug companies for many years still continued basically unchanged into the 1980s. Officials repeatedly told the newspaper that they had opted for a cooperative and "open door policy" with Canadian drug company officials instead of a tough adversarial stance. They were also proud of how friendly their relations were with representatives of Canadian drug subsidiaries of United States companies. "We try to work things out together", said one Canadian official (Regush 1982) (p. A6). Monique Bégin, the Health Minister, said that she did not believe that Canada needed detailed regulations to control the drug approval process. In her view, "the players" involved in getting a drug onto the market—federal officials, drug manufacturers and physicians conducting clinical trials—should "be forced to continue to use their heads and judgment" in carrying out the guidelines. Nor did Bégin see the need to review pre-1963 drugs for safety and effectiveness (Hollobon and Lipovenko 1982). The pharmaceutical industry was in agreement with Bégin. According to one of its representatives, Guy Beauchemin, "some companies are strongly opposed to a review of drugs, for economic reasons…A hell of a lot of products would drop out" (Regush 1982).

The same friendly cooperative attitude continued into the late 1990s when Dr. George Paterson, director-general of Health Canada's Food Directorate, was interviewed in the Globe and Mail. He espoused "the 'trust-industry' philosophy prevalent among senior managers and bureaucrats. The pharmaceutical industry and the health-protection branch [a part of Health Canada] have a 'shared purpose' to ensure no harmful products make it to market, he said. 'In terms of good business savvy, they don't want to be in any situation where their integrity and their competitiveness would be compromised by a scare'" (Eggertson, 1997) (p. 5).

The framing of perceived deficiencies in Canada's drug laws, described in a series of 5 reports dealing with the regulatory system undertaken in the 1980s and early 1990s (Auditor General of Canada, 1987; Commission of Inquiry on the Pharmaceutical Industry, 1985; Gagnon, 1992; Task Force on Program Review, 1986; Working Group on Drug Submission Review, 1987) highlighted the priority that Health Canada gave to the industry's interests versus those of the public in the policymaking arena. All of the reports conveyed a sense of urgency about the length of the drug review process, but none had any significant public input. For example, of the seven members of the Working Group on Drug Submission Review struck in the mid-1980s, six came from either government or industry. The sole other member, presumably representing the public and health professionals, came from the Canadian Public Health Association (Working Group on Drug Submission Review, 1987). In having one "outside" member the Working Group was more representative than any of the other four review teams. Although some of these reviews were entirely "in house", such as the ones from the Task Force on Program Review (Task Force on Program Review, 1986) and the Auditor General (Auditor General of Canada, 1987), the ethos that underlies these reports was strongly influenced by the clientele pluralist relationship between the government and the multinational pharmaceutical industry.

From late 2010 to early 2011, Health Canada organized a series of three public consultations to discuss the technical details of what it called regulatory modernization. The session in mid-January 2011 discussed, among other things, suspending and revoking market authorization, information requirements in advertising and establishing advisory committees. Many of the proposals were reasonable but there was also an unmistakable underlying philosophy, that the industry's needs should be privileged over those of the public. Where input was proposed with respect to suspending market authorization or questionable advertising, it was exclusively with company representatives. The documents were silent about consultations with other groups. Advisory committees would have industry representatives on them. The bias in content was mirrored by a bias in the way that the meetings were structured to reflect the nature of the relationship among Health Canada, industry and the public (Schafer, 2011). I was at one of these meetings. Officials from Health Canada sat at one long table facing another table with industry representatives. In the audience, off to the side were people from professional organizations, consumer and patient groups, many partially funded by industry, and academics. First, Health Canada would put forward proposals, then the industry would respond and only last would the people in the audience be able to ask questions or offer comments.

This section has shown how the Canadian state followed the model of clientele pluralism in looking to industry as its major partner in drug regulation. The next sections will look in detail at how the government's adoption of an alliance with industry has influenced how it regulates in the areas of clinical trials, the drug approval system, promotion, and drug safety.

5.3 Corruption in the Regulation of Clinical Trials

Before drugs can be marketed or even considered for marketing by Health Canada, they go through a series of clinical trials. By definition, clinical trials involve humans. As there are inherent risks to the people taking part, it is essential that the trials be conducted as rigorously as possible so that the data they produce can be trusted when it comes to making decisions about whether or not to approve drugs. However, there are competing interests, the industry wants trials done as quickly and as inexpensively as possible and this imperative may lead to economic values outweighing scientific ones. This section looks at how Health Canada balances these competing interests and how much trust we can place in what trials tell us about the value of new medicines.

Before any type of clinical trials on experimental drugs (drugs that have never been marketed in Canada) can proceed, they must be approved by Health Canada. Up until January 2000, the Therapeutic Products Programme (TPP), now known as the Therapeutic Products Directorate (TPD) and the Biologics and Genetic Therapies Directorate (BGTD), the branch of Health Canada dealing with prescription and nonprescription drugs, had a default time of 60 days to review applications for clinical trials. If applications had not been processed within that period, then the sponsor was free to proceed with the trial. In early 2000, Health Canada proposed to change the default time from 60 days to 48 h for Phase 1 studies. One of the main reasons offered for this change was that "the proposed option would provide the [pharmaceutical] industry with internationally competitive review times for the review of human clinical trial drug submissions" (Regulations amending the Food and Drug Regulations (1024—clinical trials), 2000) (p. 236).

In looking at changes to review times, Health Canada made it clear that it was only advancing options that would not hinder trade. Safety was definitely mentioned but seemed to take a back seat to economic considerations. What Health Canada sought to do was create conditions that would lead to increased development of the pharmaceutical industry in Canada. As a report about the benefits and costs states: "A number of firms claim to be interested in establishing facilities in Canada to conduct Phase I human clinical trials. However, it has been suggested that this can only be done if the Canadian regulatory system allows for a registration system for Phase I trials as well as reduced review times for other trials" (Regulations amending the Food and Drug Regulations (1024—clinical trials), 2000) (p. 237). While it may or may not be a reasonable objective to promote the interests of the pharmaceutical industry, this was not, and is still not, the mandate of Health Canada. The mandate of the agency was, and is, to ensure that Canadians have access to safe and effective drugs (Health Canada, 2012b). But in this case, the interests of commerce, not patients, were the explicit basis of Health Canada's policy decisions.

Companies prefer placebo-controlled trials to those using an active comparator, i.e.,, another drug that is used to treat the same condition. The logic in favour of placebo trials is simple—it's easier to show that your drug is better than nothing than to show that it's better than something else that works. Moreover, a trial using a placebo control requires fewer patients than one with an active control and the lower the number of patients the less expensive the trial is to run. Health Canada's position on the use of placebo controls is decidedly murky. According to Flood and Dyke (Flood & Dyke, 2012), Health Canada has produced a guidance document for industry on the subject (Health Products and Food Branch, 1997), but the document has no legal status as it is only intended to provide "assistance" and is not part of the Food and Drug Regulations. The guidance document is based on the International Council for Harmonization of Technical Requirements for Registration of Pharmaceuticals for Human Use (ICH) guidelines (International Conference on Harmonisation of Technical Requirements for Registration of Pharmaceuticals for Human Use, 1996), "but neither [Health Canada nor the ICH] is clear about the use of appropriate comparators; although the latter does stipulate that trials should be conducted to accord with the principles of the Declaration of Helsinki (WMA Declaration of Helsinki—ethical principles for medical research involving human subjects, 2008). The Declaration, in turn, requires that new drugs be tested against the best current proven treatment except where no current proven intervention exists or where there are compelling and scientifically sound methodological reasons to use a placebo control" (Flood & Dyke, 2012) (p. 292). Importantly, nowhere does the ICH guideline unambiguously state that an active control should be used in preference to placebo. Health Canada's silence on the question of placebo controls allows industry to continue to use them, despite their negative consequences for doctors and patients.

Health Canada conducts inspections of clinical trials and says that it uses "risk-based criteria" to select the trials to be inspected but these criteria are vague and merely state that they include the number of subjects enrolled and the number of trials conducted at a specific location (Health Products and Food Branch, 2004). Therefore, it is unknown whether Health Canada is actually monitoring the trials where the subjects are at the greatest risk. Moreover, "Health Canada does not regularly collect all of the information necessary to assess these factors and to make comparative risk-based decisions. Because clinical trial sponsors are not required to submit up-to-date information on clinical trial sites, inspectors must call each site directly to find out the current status of the clinical trial site and the number of participants enrolled…Thus, inspectors have up-to-date information only for sites that they call and are unable to compare the risks posed by all sites" (Auditor General of Canada, 2011) (p. 9). New interim risk-based criteria for trial inspections were developed and piloted by Health Canada for use during 2012–2013 in conjunction with the development of a new site selection process (Update and response to OAG recommendations for the regulation of pharmaceutical drugs in fall 2011, 2013) but there is no published information about these new criteria or how they are being applied.

Health Canada used to be much less transparent in the information it publicized about its inspections compared with the US. When the Food and Drug Administration found serious problems, the results of its inspection and its warning letter to the company were posted on its website, albeit with some delay and certain information redacted (Shuchman, 2008). In contrast, Health Canada only provided summary statistics about its inspections but no information about individual sites, corrective actions required, the drugs involved, the names of the doctors running the trials or the names of the companies (Health Products and Food Branch Inspectorate, 2012; McLean and Bruser, 2014).

In recent years Health Canada has become more proactive and now has a website (http://www.healthycanadians.gc.ca/apps/gcp-bpc/searchResult-en.html?lang=en&sponName=&contNum=&region=&drugName=&phase=&startDate=&rating=&cat=1) that posts the results of clinical trial inspections and contains information about multiple components of the inspection including whether the sponsor was compliant or noncompliant. However, if the sponsor was noncompliant there continues to be minimal information about what measures Health Canada requires. For example, in the case of AOSpine North America, the following is all that appears: "Proposed corrective actions were requested and are pending". (http://www.healthycanadians.gc.ca/apps/gcp-bpc/fullReportCard-en.html?lang=en&gcpid=4462d630-3d96-4cd1-9365-60e242af6208).

A number of high profile scandals have led to a growing call for transparency in the results of clinical research. After a series of hearings on prescription drugs in 2003, the House of Commons Standing Committee on Health recommended that Health Canada create a "…public database that provides information on trials in progress, trials abandoned and trials completed" (House of Commons Standing Committee on Health, 2004) (p. 13). By the time that Health Canada first held a public workshop about a clinical trial registry in June 2005, there were already functioning registries in Australia/New Zealand, (http://www.anzctr.org.au/Faq.aspx), the European Union (https://www.clinicaltrialsregister.eu/about.html), and the United States, (https://clinicaltrials.gov/ct2/about-site/history). Following the workshop, an external working group met in April 2006, and in June–July 2006 the public was given the opportunity to complete an online questionnaire on the topic (Health Canada, 2006).

The external working group delivered its report in December 2006 (External working group on the registration and disclosure of clinical trial information (EWG-CT), 2006). According to information on the Health Canada web site, after that report was published "Health Canada will consider the results of the public consultations and the External Working Group's recommendations before making a final decision on how to proceed with the registration and disclosure of clinical trial information in Canada" (Health Canada: Drugs & Health Products, 2007). In its 2007 Blueprint for Renewal II: Modernizing Canada's Regulatory System for Health Products and Food, Health Canada committed to enhancing public access to clinical trial information (Health Products and Food Branch, 2007a).

Finally, after almost 8 years of delay, and as part of its response to the 2011 Auditor General's report (Update and response to OAG recommendations for the regulation of pharmaceutical drugs in fall 2011, 2013), the federal government announced a public database of Health Canada-authorized clinical drug trials that will contain the titles of trial protocols, the medical conditions involved, the drugs and populations being studied, enrollment status and the dates Health Canada authorized the trials (Harper government launches clinical trials database—new initiatives provide guidance and education on clinical trials for Canadians, 2013; Shuchman, 2013). However, the Health Canada database is mainly designed to help Canadians locate ongoing clinical trials to facilitate enrollment. In contrast to other international clinical trial registries such as Clinicaltrials.gov in the US, it will not list study sites, investigators and their contact information, details of trial designs, primary and secondary outcome measures or trial results (Groves, 2008). This is the type of information that allows researchers, doctors and academics to access information about all clinical trials on new drugs, not just the trials that have been published in medical journals (Shuchman 2013).

5.4 Corruption in the Approval of Drugs

Traditionally, the money to operate the drug regulatory system has come from parliamentary appropriations, since drug regulation is seen as a service to the public. In 1985, a Task Force led by Deputy Prime Minister Eric Nielsen identified cost recovery opportunities for the premarket evaluation of drugs. Cost recovery essentially means charging the users of a service, in this case charging pharmaceutical companies for reviewing their drug applications. In 1994, the government sought to eliminate the large budgetary deficit of the early 1990s. Cost recovery for the operation of the drug approval system was introduced to compensate for the reduction in direct government funding. Cost recovery was also seen as "a means of transferring some or all of the costs of a government activity from the general taxpayer to those who more directly benefit from or who 'trigger' that special activity" (KPMG Consulting LP, 2000). Initially, pharmaceutical companies paid $143,800 to have a new drug application reviewed and by 2014 that had risen to $322,056 (Pharmaceutical submission and application review fees as of 1 April 2014, 2014). As of 2015, cost recovery was expected to provide about half of the operating budget for the various drug programs of Health Canada (Health Canada's proposal to parliament for user fees and service standards for human drugs and medical devices programs, 2010).

This shift in the financing of the regulatory body has raised concerns, based on principal–agent theory, about whether Health Canada's primary commitment continues to be to public health. Principal–agent theory proposes that there is a relationship between a principal who has a task that needs to be performed and an agent who is contracted to do the task in exchange for compensation (Evans, 1984). Prior to the introduction of user fees, the principal was the Canadian public and the agent was Health Canada. However, since 1994, a new principal has been added with: the pharmaceutical industry that is now providing a substantial fraction of the money needed to run the drug regulatory system.

One example of how cost recovery may have led to a principal–agent relationship between the industry and Health Canada was the change in how quickly drugs moved through the review process. The loudest and most influential voice calling for faster drug approvals has long come from the brand-name industry. In the past, Innovative Medicines Canada has repeatedly complained about the excessive length of time that it takes to get a drug approved (Rx&D, 2003, 2004). Industry sponsored meetings in 1986 and 1987 to publicize the issue among a variety of professional and consumer groups and put forward its message that "a supportive policy environment in terms of the regulatory review process is another factor critical to the global competitiveness of the pharmaceutical industry" (Towards a globally competitive research-based pharmaceutical sector. A submission to the steering group on the federal government's prosperity initiative 1992) (p. 16). In its 2002 report, *Improving Health Through Innovation: A New Deal for Canadians*, industry explicitly linked cost recovery fees to the "attainment of revised internationally competitive performance standards" (Rx&D, 2003).

Following the introduction of user fees, Health Canada continued to commit new revenue to ensure rapid drug review times. The budget speech in 2003 announced $190 million in new funding over 5 years to improve the timeliness of Health Canada's regulatory processes with respect to human drugs (Department of Finance, 2003). The following year, out of the $40 million allocated to Health Canada for its drugs programs for fiscal year 2003/2004, 78% ($31.2 million) went toward "improved regulatory performance"—mainly an effort to eliminate the backlog in drug approvals and to ensure timeliness in getting drugs onto the market (Health Canada, 2003b). The 2006/2007 *Departmental Performance Report* gave one of Health Canada's goals as having "90% of new pharmaceutical…drug submissions decisions met within internationally comparable performance targets, compared to 13%…in 2003" (Health Canada, 2007) (p. 40). The report briefly mentioned drug safety but no targets were set out for topics such as how quickly adverse drug reaction reports should be evaluated. Health Canada now is proposing that user fees cover 90–100% of the operating costs of the drug regulatory system (Health Canada, 2017).

Shortly after cost recovery started, Health Canada introduced its priority review pathway, intended "for a serious, life-threatening or severely debilitating disease or condition for which there is substantial evidence of clinical effectiveness that the drug provides…effective treatment, prevention or diagnosis of a disease or condition for which no drug is presently marketed in Canada; or…a significant increase in efficacy and/or a significant decrease in risk such that the overall benefit/risk profile is improved over existing therapies" (Health Canada: Health Products and Food Branch, 2009). The company seeking approval still has to submit a complete new drug submission containing the results of Phase I–III clinical trials, but the review period is reduced from the standard 300 days to 180 days. An analysis of how appropriately Health Canada uses the priority approval pathway shows that it is using this process for many drugs that do not offer any new significant benefits. From 1997 to 2012, 345 of the drugs approved by Health Canada had their therapeutic value assessed by one or two independent sources. Although Health Canada used a priority review in 91 cases, only 52 were rated as significant therapeutic advances. The positive predictive value for Health Canada's use of priority reviews, a measure of how well it accurately predicted the real value of drugs, was only 33% (Lexchin, 2015).

Another avenue for getting promising drugs to the market faster is conditional approval. In Canada, provisions for conditional approval are set out in Health Canada's Notice of Compliance with conditions (NOC/c) policy. The goal of this policy is to "provide patients suffering from serious, life threatening or severely debilitating diseases or conditions with earlier access to promising new drugs" (Health Canada, 2002). In return for having their drugs approved through the NOC/c process, companies sign a Letter of Undertaking promising to complete confirmatory clinical studies to definitively establish efficacy and submit the results of these studies to Health Canada. If these confirmatory studies do not show that the drug is efficacious Health Canada has the option of suspending the sale of the drug (Health Canada, 2002).

In at least one case, Health Canada did not suspend the sale and allowed a drug to stay on the market despite not fulfilling its conditions. Gefitinib (Iressa) was approved under this policy as a third-line treatment for non-small cell lung cancer on the condition that the company submitted a study showing that it improved survival (Health Canada, 2005). When the study results were submitted to Health Canada they showed no survival benefit for gefitinib compared to placebo (Thatcher et al., 2005). Health Canada recognized that the conditions had not been fulfilled, but rather than removing gefitinib from the market, in February 2005 it elected to allow it to continue to be sold on the rationale that "1. There is no alternative therapy available for the treatment of Canadian NSCLC patients who failed two lines of therapy; 2. Iressa® shrinks tumours, which may lead to less shortness of breath, less pain and less cough; [and] 3. The safety profile of Iressa® is more acceptable than that of any other chemotherapy which may be considered in this situation" (Health Canada, 2005) (p. 2). (In 2009, the drug was deemed to have met its conditions after a new study showed non-inferiority, i.e., survival after taking it was no worse compared to another chemotherapy drug (Kim et al., 2008).) Some drugs have been on the market for more than 10 years without fulfilling their conditions, leaving both clinicians and patients in the dark about how effective the products actually are (Law, 2014; Lexchin, 2007).

Transparency is a term that is often thrown around when it comes to drug regulation. When Health Canada uses the term, the meaning that it gives it is reassuring—"Canadians also want to better understand the decision-making process undertaken by the regulator. Health Canada must provide greater and more meaningful transparency by enabling easier access to information, as well as providing information in a format that is easy to understand and provides value to the end user" (Health Canada, 2012b) (p. 12). Understanding decision-making means two things—providing the information that was used to make the decision and disclosing in detail the reasoning process behind the decision. In practice, this means that Health Canada should make public the efficacy and safety information that it receives from drug companies and the reports from its reviewers about the quality of that information.

The Canadian Association of Journalists deemed Health Canada the most secretive of all government departments and in 2004 awarded it its fourth annual "code of silence" award for showing "remarkable zeal in suppressing information" and "concealing vital data about dangerous drugs" (Kermode-Scott, 2004). In early September 2011, Toronto doctor Nav Persaud applied to Health Canada for all of the information that it had on Diclectin (doxylamine and pyridoxine), a widely used drug for morning sickness. Six months later, Health Canada told him that it was "negotiating with the third party [Duchesnay, the maker of the drug] on which records they wish us to protect". Finally, after more than a year, he received 359 pages of which 212 were completely censored. "Other pages had blacked-out sections under titles such as 'Adverse Events,' because they were deemed confidential business information" (Bruser et al. 2015).

The main rationale used by Health Canada to support secrecy is that it regards virtually all information that it receives from a drug company as commercially confidential that cannot be released without the expressed permission of the company. Not surprisingly, companies typically invoke their right to withhold information, arguing that if this information became public knowledge it would harm their economic interest. This line of reasoning was reflected in a 2004 article authored by a vice-president of Merck, one of the largest pharmaceutical companies in the world. Writing in the *CMAJ*, he stated that "Merck (like other companies) is obliged to protect proprietary information and intellectual property, including aspects of the design of clinical trials of investigational agents and the very existence of certain studies" (Hirsch, 2004) (p. 482). Health Canada "considers it inappropriate to post information concerning clinical trials, special access or products with notices of compliance with conditions on the basis that this information is incomplete or open to misinterpretation" (The Public Policy Forum, 2003) (p. 8).

A second argument against transparency is based on Health Canada's reading of the requirements in the free trade agreements that Canada has signed, specifically the North American Free Trade Agreement and the Trade-Related Aspects of Intellectual Property Rights Agreement. According to a Health Canada proposal for legislative renewal, these agreements "provide that the government must protect the confidentiality of undisclosed test or other data provided by the applicant to determine whether the use of such products is safe and effective, where the origination of such data involves considerable effort". In light of this interpretation, during one set of consultations on health protection legislation renewal in 2003, Health Canada was only willing to propose that a summary of the safety and effectiveness data submitted by the manufacturer would be made generally public. If consumers or health professionals want to see the full set of data, the proposal called for a "reading room where people could review all the data submitted by the manufacturer but not transcribe or copy it, or otherwise make that data available to interested members of the public" (Health Canada, 2003a) (p. 76).

Matthew Herder, who teaches law at Dalhousie University in Halifax, commented on how these rationales from Health Canada played out in the real world based on his participation in a series of technical meetings between October 2010 and January 2011 held to discuss regulatory modernization at Health Canada. "Each proposal put on the table by Health Canada to increase transparency—from making final decisions regarding applications for market authorization publicly available, to creating an online register of therapeutic products – was met with proprietary claims from MEDEC [Canada's Medical Technology Companies], BIOTECanada [the national biotechnology industry association] or Rx&D [now Innovative Medicines Canada], the respective associations of medical device, biotechnology and pharmaceutical companies in Canada. Each time, Health Canada acknowledged that the law controlled what they could and could not disclose" (Herder, 2012) (p. 194) and "the moderator noted the issue on a flip chart dubbed the "parking lot" and urged participants to stick to the agenda" (Herder, 2012) (p. 198).

In March 2017, Health Canada released a discussion paper about making clinical information public (Health Canada: Health Products and Food Branch, 2017) and followed this up with a posting in Canada Gazette (Regulations amending the Food and Drug Regulations (Public Release of Clinical Information): regulatory impact analysis statement 2017). Based on the contents of these two documents, Health Canada is proposing to release nearly all of the clinical data that drug companies submit when they want to gain approval to market a new drug. If Health Canada follows up on this commitment, and at the time of writing (April 2018) there is every indication that it will do so, this will represent a major commitment to transparency.

5.5 Corruption in the Regulation of Promotion

The F&D Act and Regulations give the government authority over the promotion of both prescription and over-the-counter drugs and ban direct-to-consumer advertising (DTCA). But despite having the nominal control over promotion, over the past century Health Canada and its predecessors have never shown much interest in exercising that power. Health Canada's position is that "it's not our policy to treat advertising as the definitive source of information with respect to drugs" (Cocking, 1977) (p. 19) and therefore, it only makes informal spot checks on drug advertising. No penalties were imposed on any pharmaceutical company for illegal advertising between 1978 and 1984 (Lexchin, 1984) and there is no public record of any since then. The Regulatory Advertising Section within the Marketed Health Products Directorate

(MHPD), the part of Health Canada charged with regulating drugs (and other health products) already on the market, oversees regulated advertising activities but the exact nature of the Regulatory Advertising Section's activities is opaque. Despite an extensive search on Health Canada's website there is no information about the number of personnel in the Section or its level of resources.

In the face of egregious examples of medical journal advertising featuring, among other images, nude and seminude women, a 1973 meeting of federal and provincial health ministers recommended that the federal government "review controls on the advertising of drugs with the aim of strengthening them where necessary" (Raison, 1989) (p. 27). In response, Marc Lalonde, then the federal Minister of Health and Welfare, issued an ultimatum to the industry to reform its practices or else face the prospect of government action. Pharmaceutical Manufacturers Association of Canada (PMAC, now Innovative Medicines Canada) was very much in favour of self-regulation over direct control by the government and initiated a sequence of events that resulted in the creation of the Pharmaceutical Advertising Advisory Board (PAAB) in 1975. The first chair of the PAAB board was also an employee of the Upjohn company and the PMAC promotion code formed the basis for the one adopted by PAAB (Raison, 1989). PAAB was given the responsibility for advertising/promotion systems, i.e., the media presentation of promotion in all forms—print, audio, visual, audio/visual, and later electronic and computer means of communication. Advertisements in any of these forms have to be submitted to the PAAB for preclearance to ensure compliance with the provisions of its code before they can be used (Pharmaceutical Adverising Advisory Board, 2013). The industry code retained responsibility for all other forms of promotion including sales representatives, gift-giving, and the sponsorship of meetings. The industry code also mandates compliance with that of the PAAB (Innovative Medicines Canada, 2016).

Despite the nominal independence of the PAAB from industry, its board includes a representative of the Association of Medical Advertising Agencies, the Canadian Association of Medical Publishers, Canadian Generic Pharmaceutical Association, Consumer Health Products Canada, IMC and BIOTECanada, meaning that 6 of the 14 members come from organizations that in one way or another benefit from advertising (Pharmaceutical Advertising Advisory Board, 1986). Since it takes a two-thirds majority of the board to make changes to the code it is highly unlikely that anything that is unacceptable to the industry will ever be passed. The PAAB's code is not legally binding; its decisions are not legally enforceable and as a voluntary, independent body, the PAAB is not accountable for its actions to the government or any other organization. PAAB will occasionally refer complaints about ads to Health Canada for final resolution. This method of regulating advertising is apparently acceptable to Health Canada, as it is an ex-officio member of the board, demonstrating again the clientele pluralist relationship between Health Canada and industry.

There are a number of major weakness in the PAAB code, one of which is that it does not require any information in ads that does not appear in the Product Monograph (PM), the scientific document on a drug product that, devoid of promotional material, describes the properties, claims, indications and conditions of use of the drug and contains any other information that may be required for optimal, safe and effective use of the drug, and is approved by Health Canada. The use of surrogate outcomes in ads shows the problem with this limitation. Many drugs are approved by Health Canada solely on the basis of surrogate outcomes. A surrogate outcome could be a change in tumour size, in a physiologic measurement such as blood pressure or in a laboratory value. However, surrogate outcomes do not necessarily translate into the things that matter for patients, a change in morbidity and/or mortality, i.e., a hard clinical outcome. There are a number of instances where drugs approved on the basis of surrogate outcomes later had to be pulled from the market because they did more harm than good. For example, Enkaid (encainide) was initially approved because it suppressed irregular heartbeats but was eventually found to increase overall mortality (Svensson et al., 2013). If the PM lacks a statement saying that the medication has not been shown to reduce morbidity and/or mortality, then advertisements are not required to contain a statement to this effect. The ads will just mention the changes in surrogate outcomes. However, because doctors rely so heavily on promotion, the absence of a statement about the lack of changes in morbidity and/or mortality can be a problem. Doctors may believe that claims about positive changes in surrogate outcomes translate into a reduction in morbidity and/or mortality.

Not surprisingly, given the composition of the PAAB's board, there continue to be other significant problems with its code. While the code puts a lower limit on the font size for print there is no requirement to make the generic name of the drug the same size as the brand name and additionally, the generic name does not have to be used each time that the brand name is given despite evidence that use of the generic name leads to better prescribing (Spurling et al., 2010). Companies are allowed to make statements in journal ads about the effects of drugs, even if the clinical significance of those effects is unknown, as long as the ad also includes that caveat. Finally, and maybe most significantly, there are no significant sanctions for violating the PAAB code.

Over 50% of the amount that companies spend on promotion goes towards the expenses of their salespeople, the men and women who visit doctors in their offices and clinics and hospitals (Gagnon & Lexchin, 2008). The Code of Ethical Practices from Innovative Medicines Canada requires company sales representatives to "provide accurate and up-to-date information", presumably including information about how to safely use the product they are promoting. This provision in the code is reinforced in a statement about the value of detailing: "Conversations between health care providers and representatives of Canada's research-based pharmaceutical companies focus on the appropriate use of medicines" (Rx&D, 2010). Whether the representatives do this in practice was recently investigated by having general practitioners in Montreal and Vancouver fill in survey forms after they had seen a representative. The primary outcome measure was presence of "minimally adequate safety information", defined a priori as the mention of $\geq$1 approved indication, $\geq$1 serious adverse event, $\geq$1 common non-serious adverse event, $\geq$1 contraindication *and* no unapproved indications or unqualified safety claims (e.g., "this drug is safe"). "Minimally adequate safety information" was provided in 5/412 (1.2%) of promotions in Vancouver and 7/423 (1.7%) in Montreal. Representatives did not provide any information about harms (a serious adverse event, a common adverse event or a contraindication) in 66% of interactions (Mintzes et al., 2013).

Over the past two decades, Health Canada has become increasingly tolerant of DTCA, despite the prohibition of this type of advertising in the Food and Drug Regulations (2014). Barbara Mintzes, who teaches pharmaceutical policy at the University of Sydney, and I have examined how well Health Canada is regulating DTCA (Lexchin & Mintzes, 2014). To do this we used a case study approach and looked at 10 examples of DTCA, involving 8 different drugs, that appeared to contravene Health Canada's policy on DTCA. Our overall conclusion was that Health Canada had adopted a narrow approach to enforcement and ignored broader concerns such as off-label promotion, targeting of vulnerable groups, and poor safety profiles of products. Only one enforcement tool was used, negotiation with the company; fines, sanctions, requirements for remedial action or prosecutions were not used.

5.6 Corruption in the Monitoring of Drug Safety

Between 1990 and 2009, about 4–5% of all new drugs approved by Health Canada in 4 different 5-year periods (1990–1994, 1995–1999, 2000–2004, 2005–2009) subsequently needed to be withdrawn from the market for safety reasons (Lexchin, 2013). Although the percentage of withdrawals has remained relatively stable, aggressive marketing of new drugs by companies means that an increasing number of people may be exposed to these products before they are removed from the market. Despite significant safety concerns, there is evidence that Health Canada is not monitoring safety adequately. Two of the 5 most heavily promoted drugs in Canada in 2000 Vioxx (rofecoxib), used for pain and inflammation, and Baycol (cerivastatin), used for high cholesterol were subsequently withdrawn because of safety issues (Lexchin, 2013). Surprisingly, Health Canada does not keep a specific summary record of drugs that have been withdrawn for safety reasons. As a result, acquiring such information requires hand searching multiple Health Canada databases. The absence of such a list means that Health Canada cannot easily look at or track trends in drug withdrawals; it cannot examine how long it takes to identify serious safety problems and whether this situation is worsening or improving, and it cannot see if some manufacturers have more drugs with safety problems than others.

Industry maintains that its first priority is the health and well-being of Canadians (Innovative Medicines Canada, 2016) and likewise "Health Canada's vision is to continually promote high standards of product vigilance for the protection of the health and safety of Canadians" (Health Canada, 2012a) (p. 1) but Health Canada's actions belie this statement. Health Canada documents about what triggers a safety action are quite vague and provide little transparency or specifics regarding how decisions are made. For example, Health Canada states "The determination of the seriousness of risk (probability of health hazard and probability of occurrence) and urgency of risk communication is based on sound scientific judgement" (Draft guidance document—triggers for issuance of risk communication documents for marketed health products for human use 2007). "Regulatory actions…are taken according to the regulatory framework in place. This implies an evaluation of the signal and the appropriate benefit–risk review of the information available" (Marketed Health Products Directorate, 2004) (p. 6).

Chris Turner, the previous head of the MHPD, wrote a letter to the Toronto Star defending Health Canada's drug safety program, saying "Health Canada has highly trained specialists who use Canadian adverse reaction data as well as other sources of information to systematically monitor, analyze and act on safety issues" (Turner, 2012). But another, unnamed, Health Canada official is quoted as telling the Toronto Star "It is primarily the [drug company's] responsibility to monitor the safe use of their products" (Bruser & Bailey, 2012).

In May 2005, Christiane Vellemure, the Director of the Office of Business Transformation, Planning and Administration at Health Canada showed a Powerpoint presentation looking at the progress that had been made in setting up a new Health Product Safety Board that was designed to "encourage public input in regulatory decision making, strengthen public accountability in regulation and risk management of health products [and] provide independent advice on health product safety issues as a permanent part of the Health Canada decision making process" (Villemure, 2005). By the time this body had its first meeting in November 2007, it had morphed into the Expert Advisory Committee on the Vigilance of Health Products with a mandate to provide objective external expert advice on broad strategic policy and program issues involving marketed therapeutic health products for human use (Health Products and Food Branch, 2007c). The committee met 2–3 times per year until 2011 (Committee Meeting Summary Reports, 2012), but since then there have been no further meetings despite an announcement on the web site of a meeting in October 2012 (Expert Advisory Committee on the Vigilance of Health Products, 2012). According to one anonymous member of the committee, it advised the Minister of Health only a single time. This same member recounts that the committee asked Health Canada to define how safety signals were determined and what criteria or weighting was used in deciding about communicating safety problems. The committee never heard back from Health Canada (Committee member, personal communications, 24, October 2010 and 11 January 2015).

Health Canada places safety at a lower priority than its mandate to get new drugs onto the market. In 2004, the arms of Health Canada that approved drugs (Therapeutic Products Directorate (TPD) and Biologics and Genetic Therapies Directorate (BGTD)) were getting about seven times more funding and had about seven times more personnel than the MHPD. The 2004 imbalance did not seem to be a priority for the government. Out of $40 million allocated to Health Canada in 2003–2004 for its drugs programs, just $2.5 million went to monitoring postmarket drug safety (Health Canada, 2003c). A 2006 investigation by the Auditor General documented that drug program managers felt that they were unable to meet the regulatory requirements for monitoring the safety of marketed drugs because of a lack of funds (Auditor General of Canada, 2006). Although the situation was marginally better by the end of March 2017, there was still more than three times the number of personnel and amount of funding going to the directorates reviewing drug applications (TPD and BGTD) compared to the one charged with monitoring safety (MHPD) (Lexchin, unpublished data).

This imbalance in personnel and resources at Health Canada is also reflected in a power imbalance. While the MHPD monitors safety, it does not have the authority to send out safety warnings or to withdraw products from the market. Those decisions remain with the TPD and the BGTD, the parts of the agency that authorized the marketing of the drugs and, therefore, in order to take action these directorates have to be willing to admit that they may have made a mistake in the first place. (It should be noted, of course, that not all drug safety problems can be anticipated at the time that drugs are being considered for marketing.) Whether this division of authority leads to tensions within Health Canada has not been explored.

In an effort to ensure that promising therapies for serious illnesses can reach Canadians in a timely manner, Health Canada has developed two pathways for approving new drugs more rapidly—priority reviews and the Notice of Compliance with conditions (NOC/c) policy. A priority review reduces the review period from the standard 300 days to 180 days. If companies apply for NOC/c status when they file the new drug submission and Health Canada agrees to the NOC/c application then drugs are reviewed in 200 days. If companies do not initially apply for NOC/c status, then drugs are reviewed in either 180 or 300 days and Health Canada may grant NOC/c status at the end of the review.

An important question to be answered, is whether the shorter priority review period and the approval of drugs under the NOC/c policy, leads to these products having more serious safety issues once they are marketed. The answer seems to be yes. If a drug has a standard review, there was a 20% chance that Health Canada will either issue a serious safety warning about it or else the drug will be withdrawn from the market because it was unsafe. For drugs approved with a priority review, that figure rises to 34% (Lexchin, 2012). A similar situation exists for drugs approved through the NOC/c policy. In that case, it was just under 41% of drugs that either had a serious safety warning or were withdrawn (Lexchin, 2014). For both NOC/c and priority review drugs, the increased safety problems weren't balanced by an increase in benefits.

Closely related to the speed of drug reviews is the issue of tying review times to the amount that the regulatory agency receives in user fees, as is done in the US and Canada. As the deadline approaches, reviewers may inadvertently speed up their examination of the evidence and overlook safety issues in order to avoid a loss in revenue. Carpenter and colleagues examined this question in the US (Carpenter, 2008; Carpenter et al., 2008) where the FDA has a statutory requirement to complete its review of 90% of new drug applications within set periods of time. If the FDA fails to meet that obligation, then the renewal of legislation that allows it to collect user fees from industry may be endangered. The conclusion reached was that when drugs are approved in the immediate pre-deadline period (within 2 months of the deadline), there is a substantially higher rate of withdrawals and/or safety labeling changes compared to drugs approved when the deadline was not an issue.

Similarly, cost recovery fees paid to Health Canada also suffer if reviews of new drug applications are not completed within the targeted time. If Health Canada takes too long to review applications for new drugs in one year, then the next year it is forced to reduce user fees. For example, if average review times are 20 percent over time in 2012, then fees will drop by 20 percent in 2013 (Health Products and Food Branch, 2007b). Health Canada's new proposal to increase the percentage of operating costs covered by user fees also has a provision for individual company applications. If these are not completed within the established performance standard, 25% of the fee would be rebated (Health Canada, 2017).

Faced with the prospect of penalties, it is possible that Health Canada might follow the pattern set by the FDA and perhaps rush to approve new drugs that are approaching the deadline in order to avoid incurring a financial loss in the next year. The equivalent would be putting a time limit on how long the Transportation Safety Board had to investigate an airplane crash and if it exceeded that limit its funding would be cut. At the time that this chapter is being written (April 2018), there is no evidence about whether this new policy has had any effect on drug safety.

The most recent piece of legislation to try and deal with the question of postmarket safety is Bill-17 that was signed into law in November 2014 (Herder et al., 2014). Once the regulations for Bill C-17 are written and the legislation comes into force it will correct the flaw in the current Food and Drugs Act whereby only the manufacturer can actually remove a drug from the shelves. Bill C-17 gives the Minister the explicit authority to issue a recall, without prior consultation with the manufacturer, if the Minister "believes that a therapeutic product presents a serious or imminent risk of injury to health" (House of Commons of Canada, 2013). Bill C-17 also gives the Minister the power to require postmarket studies when a Notice of Compliance is issued. In a potentially very positive step, Bill C-17 contains provisions to enhance the transparency of clinical trial information by giving the Minister the power to release such information where the risk of injury is suspected or for the protection and promotion of health. In addition, the Minister is given the power to make regulations that require companies to make "prescribed information" transparent and the regulations, once written, could extend the definition of prescribed information to postmarket studies.

Innovative Medicines Canada was quite clear that C-17 should only give Health Canada limited power to unilaterally recall drugs. Testifying before the Standing Senate Committee on Social Affairs, Science and Technology, Walter Robinson, the vice-president for government affairs at the organization, stated that any disclosure of information must be done in a way to protect "Confidential Business Information" which industry defined as "any information that has economic value to a business or its competitors and that is not usually publicly available" (Rhines & Robinson, 2014). The only exception acceptable to the industry would be where there is an imminent and serious threat to human health. Industry proposed amendments at the Senate hearings that would have considerably narrowed the reasons for disclosure, instead of allowing disclosure "if the Minister believes that the product may present a serious risk of injury to human health" (House of Commons of Canada, 2014). The industry also proposed changing "if the purpose of the disclosure is related to the promotion or protection of human health" to read "if the purpose of the disclosure is necessary for the protection of human health" (Rhines & Robinson, 2014). Neither those proposals from industry nor a number of amendments proposed by those who wanted to strengthen the bill by, for example, replacing the discretionary word "may" with the more prescriptive term "shall" (Standing Senate Committee on Social Affairs Science and Technology, 2014), were approved by the Senate.

5.7 Conclusion

Private values are antithetical to democracy; they speak to the need to earn a profit not to protect public health. While the two can at times be synonymous that happens mostly by coincidence rather than by design. Within the private sector, competition and the profit motive may be the best way to get newer and better computers or washing detergent. However, medications are not ordinary consumer products and government is intimately and necessarily involved with almost all aspects of medications because of their importance in health care. When the government adopts the values of private industry in drug regulation it is, in essence, telling its people that the needs and values of the private sector take precedence over their health. Democracy is not just the right to vote in an election it means the ongoing and active participation of the citizenry in determining the policies of the government with an expectation that the government will acknowledge the views being put forward and incorporate them into its actions. This chapter has shown that the relationship between Health Canada and industry has corrupted the regulatory system and as a result, Health Canada has largely decided to ignore those democratic values opting instead for a drug regulatory system that reflect the interests of private industry.

References

Anderson, J. (1977). Pressure groups and the Canadian bureaucracy. In K. Kernaghan (Ed.), *Public administration in Canada: Selected readings* (3rd ed., pp. 292–304). Toronto: Methuen.

Atkinson, M. M., & Coleman, W. D. (1985). Corporatism and industrial policy. In A. Cawson (Ed.), *Organized interests and the state* (pp. 22–44). London: Sage.

Atkinson, M. M., & Coleman, W. D. (1989). *The state, business, and industrial change in Canada*. Toronto: University of Toronto Press.

Auditor General of Canada. (1987). Report to the House of Commons, fiscal year ended 31 March 1987. Ottawa.

Auditor General of Canada. (2006). Report of the Auditor General of Canada to the House of Commons. Chapter 8: allocating funds to regulatory programs - Health Canada. Ottawa: Office of the Auditor General of Canada.

Auditor General of Canada. (2011). Chapter 4: Regulating pharmaceutical drugs - Health Canada. Ottawa.

ADHD drugs suspected of hurting Canadian kids. (2012, September 26). *Toronto Star*.

Begin plans no big changes in regulations for new drugs. (1982, October 28). *Globe and Mail*, 5.

Bruser, D., McLean, J., & Mendleson, R. (2015, April 24). Toronto doctor asks Health Canada about pregnancy drug, gets 212 pages of censored information. *Toronto Star*.

Carpenter, D. (2008). Drug-review deadlines and safety problems. *New England Journal of Medicine, 359,* 96–98.
Carpenter, D., Zucker, E. J., & Avorn, J. (2008). Drug-review deadlines and safety problems. *New England Journal of Medicine, 358,* 1354–1361.
Cocking, C. (1977, June 18). The abuse of prescription drugs. *Weekend Magazine,* 16–19.
Commission of Inquiry on the Pharmaceutical Industry. (1985). Report. Ottawa.
Committee Meeting Summary Reports. (2012). http://www.hc-sc.gc.ca/dhp-mps/medeff/advise-consult/eacvhp-ccevps/meet-reunion/index-eng.php. Accessed September 17 2014.
Department of Finance. (2003). Building the Canada we want. Budget 2003: Investing in Canada's health care system. Retrieved February 15, 2004, from http://www.fin.gc.ca/budget03/booklets/bkheae.htm.
Draft Guidance Document—Triggers for Issuance of Risk Communication Documents for Marketed Health Products for Human Use. (2007). Retrieved September 26, 2013, from http://publications.gc.ca/collections/collection_2007/hc-sc/H164-48-2007E.pdf.
Drug-testing rules broken by Canadian researchers. (2014, September 16). *Toronto Star.*
Eggertson, L. (1997, May 28). Drug-approval process criticized. *Globe and mail* (pp. A1, A10).
Evans, R. (1984). *Strained mercy: The economics of Canadian health care.* Toronto: Butterworths.
Expert Advisory Committee on the Vigilance of Health Products. (2012). Retrieved September 17, 2014, from http://www.hc-sc.gc.ca/dhp-mps/medeff/advise-consult/eacvhp-ccevps/index-eng.php.
External Working Group on the Registration and Disclosure of Clinical Trial Information (EWG-CT). (2006). Final report: "Options for improving public access to information on clinical trials of health problems in Canada". Ottawa: Health Canada.
Flood, C., & Dyke, P. (2012). The data divide: Managing the misalignment in Canada's evidentiary requirements for drug regulation and funding. *UBC Law Review, 45,* 282–325.
Gagnon, D. (1992). Working in partnerships … drug review for the future. Ottawa.
Gagnon, M.-A., & Lexchin, J. (2008). The cost of pushing pills: a new estimate of pharmaceutical promotion expenditures in the United States. *PLoS Medicine, 5*(1), e1.
Groves, T. (2008). Mandatory disclosure of trial results for drugs and devices. *BMJ, 336,* 170.
Harper Government Launches Clinical Trials Database—New Initiatives Provide Guidance and Education on Clinical Trials for Canadians (2013). Retrieved September 7, 2014, from http://www.hc-sc.gc.ca/ahc-asc/media/nr-cp/_2013/2013-70-eng.php.
Health Canada. (2002). Notice of compliance with conditions (NOC/c). Ottawa.
Health Canada. (2003a). Health protective legislative renewal: Detailed legislative proposal. Ottawa.
Health Canada. (2003b). Improving Canada's regulatory process for therapeutic products: Building the action plan: Multistakeholder consultation: Public Policy Forum.
Health Canada. (2003c). Improving Canada's regulatory process for therapeutic products: building the action plan: Multistakeholder consultation: Public Policy Forum.
Health Canada. (2005). Clarification from Health Canada regarding the status of Iressa® (gefitinib) in Canada. Retrieved October 20, 2014, from http://www.hc-sc.gc.ca/dhp-mps/prodpharma/activit/fs-fi/fact_iressa-eng.php.
Health Canada. (2006). Clinical trials: registration and disclosure of information. Retrieved December 21, 2006, from http://www.hc-sc.gc.ca/dhp-mps/prodpharma/activit/proj/enreg-clini-info/2006-consult/index_e.html.
Health Canada. (2007). Departmental performance report for the period ending. Ottawa.
Health Canada. (2012a). Health product vigilance framework. Retrieved September 17, 2014, from http://www.hc-sc.gc.ca/dhp-mps/pubs/medeff/_fs-if/2012-hpvf-cvps/index-eng.php.
Health Canada. (2012b). Regulatory roadmap for health products and food. Retrieved February 15, 2015, from http://www.hc-sc.gc.ca/ahc-asc/activit/strateg/mod/roadmap-feuillederoute/rm-fr-eng.php.
Health Canada. (2017). Fee proposal for drugs and medical devices (for consultation).

Health Canada: Drugs & Health Products. (2007). Clinical trials: Registration and disclosure of information. Retrieved August 13, 2007, from http://www.hc-sc.gc.ca/dhp-mps/prodpharma/activit/proj/enreg-clini-info/2005-consult/index_e.html.

Health Canada: Health Products and Food Branch. (2009). Guidance for industry: Priority review of drug submissions.

Health Canada: Health Products and Food Branch. (2017). Public release of clinical information in drug submissions and medical device applications. Ottawa.

Health Canada's Proposal to Parliament for User Fees and Service Standards for Human Drugs and Medical Devices Programs. (2010). Retrieved September 8, 2014, from http://www.hc-sc.gc.ca/dhp-mps/finance/costs-couts/fee-propo-frais-eng.php.

Health Products and Food Branch. (1997). *Guidance for industry: Good clinical practice: consolidated guideline ICH topic E6*. Ottawa: Health Canada.

Health Products and Food Branch. (2004). Summary report of the inspections of clinical trials conducted in 2003/2004. Ottawa: Health Canada.

Health Products and Food Branch. (2007a). Blueprint for renewal II: Modernizing Canada's regulatory system for health products and food. Ottawa.

Health Products and Food Branch. (2007b). Cost recovery framework: Official notice of fee proposal for human drugs and medical devices. Health Canada.

Health Products and Food Branch. (2007c). Inaugral summary meeting report—Expert Advisory Committee on the Vigilance of Health Products—November 27–28, 2007. Retrieved September 17, 2014, from http://www.hc-sc.gc.ca/dhp-mps/medeff/advise-consult/eacvhp-ccevps/meet-reunion/nov-nov-07-eng.php.

Health Products and Food Branch Inspectorate. (2012). Summary report of inspections of clinical trials conducted from April 2004 to March 2011.

Herder, M. (2012). Unlocking Health Canada's cache of trade secrets: Mandatory disclosure of clinical trial results. *CMAJ, 184,* 194–199.

Herder, M. (2014). Denaturalizing transparency in drug regulation. *McGill Journal of Law and Health, 8*(2), S57–S143.

Herder, M., Gibson, E., Graham, J., Lexchin, J., & Mintzes, B. (2014). Regulating prescription drugs for patient safety: Does Bill C-17 go far enough? *CMAJ, 186,* E287–E292.

Hirsch, L. (2004). Randomized clinical trials: What gets published, and when? *CMAJ, 170,* 481–483.

House of Commons Debates: Official Report. (1962). Ottawa: Queen's Printer.

House of Commons of Canada. (2013). Bill C-17: An act to amend the Food and Drugs Act, first reading, December 6, 2013.

House of Commons of Canada. (2014). Bill C-17: An act to amend the Food and Drugs Act, as passed by the House of Commons June 16, 2014.

House of Commons Standing Committee on Health. (2004). Opening the medicine cabinet: First report on health aspects of prescription drugs. Ottawa.

How a suspect arthritis drug evaded government checks. (1982, October 25). *Montreal Gazette,* A-1, A-6.

Innovative Medicines Canada. (2016). Code of ethical practices. Retrieved October 18, 2016, from http://innovativemedicines.ca/wp-content/uploads/2015/06/IMC_Code_EN.pdf.

International Conference on Harmonisation of Technical Requirements for Registration of Pharmaceuticals for Human Use. (1996). ICH harmonised tripartite guideline: Guideline for good clinical practice E6(R1). Retrieved September 6, 2014, from http://www.ich.org/fileadmin/Public_Web_Site/ICH_Products/Guidelines/Efficacy/E6/E6_R1_Guideline.pdf.

Kermode-Scott, B. (2004). Canadian health ministry faces criticism for its secrecy. *BMJ, 328,* 1222.

Kim, E., Hirsh, V., Mok, T., Socinski, M., Gervais, R., Wu, Y.-L., et al. (2008). Gefitinib versus docetaxel in previously treated non-small-cell lung cancer (INTEREST): A randomised phase III trial. *Lancet, 372,* 1809–1818.

KPMG Consulting LP. (2000). Report volume 1: Review of the Therapeutic Products Programme cost recovery initiative. Ottawa.

Law, M. (2014). The characteristics and fulfillment of conditional prescription drug approvals in Canada. *Health Policy, 116,* 154–161.
Lexchin, J. (1984). *The real pushers: A critical analysis of the Canadian drug industry*. Vancouver: New Star Books.
Lexchin, J. (2007). Notice of compliance with conditions: A policy in limbo. *Healthcare Policy, 2,* 114–122.
Lexchin, J. (2012). New drugs and safety: What happened to new active substances approved in Canada between 1995 and 2010? *Archive of Internal Medicine, 172,* 1680–1681.
Lexchin, J. (2013). How safe are new drugs? Market withdrawal of drugs approved in Canada between 1990 and 2009. *Open Medicine, 8,* e14–e19.
Lexchin, J. (2014). Postmarket safety warnings for drugs approved in Canada under the Notice of Compliance with conditions policy. *British Journal of Clinical Pharmacology*, *79*, 847–859.
Lexchin, J. (2015). Health Canada's use of its priority review process for new drugs: A cohort study. *British Medical Journal Open, 5,* e006816.
Lexchin, J., & Mintzes, B. (2014). A compromise too far: A review of Canadian cases of direct-to-consumer advertising regulation. *International Journal of Risk and Safety in Medicine, 26,* 213–225.
Marketed Health Products Directorate. (2004). How adverse reaction information on health products is used. Ottawa.
Mintzes, B., Lexchin, J., Sutherland, J., Beaulieu, M.-D., Wilkes, M., Durrieu, G., et al. (2013). Pharmaceutical sales representatives and patient safety: A comparative prospective study of information quality in Canada, France and the United States. *Journal of General Internal Medicine, 28,* 1368–1375.
Morrision, A. (1975). The Canadian approach to food and drug regulations. *Food, Drug, Cosmetic Law Journal, 30,* 632–643.
Pharmaceutical Adverising Advisory Board. (1986). PAAB regulations amending the food and drug regulations (Public Release of Clinical Information): Regulatory impact analysis statement code of advertising acceptance. Pickering: PAAB.
Pharmaceutical Advertising Advisory Board. (2013). Code of advertising acceptance. Retrieved September 13, 2014, from http://www.paab.ca/paab-code.htm.
Pharmaceutical submission and application review fees as of April 1, 2014. (2014). Retrieved September 8, 2014, from http://www.hc-sc.gc.ca/dhp-mps/finance/fees-frais/pharma-eng.php.
Potter, R. (1966). Prescription drugs: The changing response of Canada's Food and Drug Act (Part 4). *Canadian Pharmaceutical Journal, 99*(2), 9–12, 39.
Raison, A. (1989). *The evolution of standards for pharmaceutical advertising in Canada*. Pickering: Pharmaceutical Advertising Advisory Board.
Regulations Amending the Food and Drug Regulations (Public Release of Clinical Information): regulatory impact analysis statement. (2017). *Canada Gazette, 151*(49).
Regulations Amending the Food and Drugs Regulations (1024—Clinical trials). (2000). *Canada Gazette, 134*(4), 227–260.
Re: ADHD drugs suspected of hurting Canadian kids, Sept. 26. (2012, September 27). *Toronto Star*.
Rhines, J., & Robinson, W. (2014). Standing Senate Committee on Social Affairs, Science and Technology, Bill C-17: Evidence. Retrieved October 19, 2014, from http://www.parl.gc.ca/HousePublications/Publication.aspx?Language=E&Mode=1&DocId=6676418.
Rx&D. (2003). Improving health through innovation: A new deal for Canadians. Ottawa.
Rx&D. (2004). Drug approval times…a question of access for Canadian patients to new and innovative therapies. Ottawa.
Rx&D (2010). Where we stand: detailing. (pp. 2). Ottawa: Canada's Research-Based Pharmaceutical Companies
Schafer, A. (2011, January 19). A bitter pill: Consultations on drug safety appear to be stacked in favour of industry, which could be bad news for the public, writes Arthur Schafer. *Ottawa Citizen*.
Shuchman, M. (2008). Clinical trials regulation—How Canada compares. *CMAJ, 179,* 635–638.

Shuchman, M. (2013). Health Canada's new clinical trials database should be mandatory, says expert. *CMAJ, 185,* 946.

Spurling, G., Mansfield, P. R., Montgomery, B., Lexchin, J., Doust, J., Othman, N., et al. (2010). Information from pharmaceutical companies and the quality, quantity, and cost of physicians' prescribing: a systematic review. *PLoS Medicine, 7,* e1000352.

Standing Senate Committee on Social Affairs Science and Technology. (2014). Evidence. Retrieved October 27, 2014, from http://www.parl.gc.ca/content/sen/committee/412/SOCI/51634-E.HTM.

Svensson, S., Menkes, D., & Lexchin, J. (2013). Surrogate outcomes in clinical trials: A cautionary tale. *JAMA Internal Medicine, 173,* 611–612.

Task Force on Program Review. (1986). Improved program delivery: Health and sports. Ottawa.

Thatcher, N., Chang, A., Parikh, P., Pereira, J., Ciuleanu, T., von Pawel, J., et al. (2005). Gefitinib plus best supportive care in previous treated patients with refractory advanced non-small-cell lung cancer: results from a randomised, placebo-controlled, multicentre study (Iressa Survival Evaluation in Lund Cancer). *Lancet, 366,* 1527–1537.

The Public Policy Forum. (2003). Improving Canada's regulatory process for therapeutic products.

Towards a globally competitive research-based pharmaceutical sector. A submission to the steering group on the federal government's prosperity initiative (1992). Ottawa.

Turner, C. (2012, September 27). ADHD drugs suspected of hurting canadian kids. *Toronto Star.*

Update and response to OAG recommendations for the regulation of pharmaceutical drugs in fall 2011. (2013). Retrieved September 9, 2014, from http://www.hc-sc.gc.ca/ahc-asc/pubs/hpfb-dgpsa/oag-bvg-eng.php.

Villemure, C. (2005). Health product safety board: Progress update: presentation to ACM.

WMA Declaration of Helsinki—Ethical principles for medical research involving human subjects (2008). Retrieved September 15, 2013, from http://www.wma.net/en/30publications/10policies/b3/.

Working Group on Drug Submission Review. (1987). Memorandum to the Minister (the Stein Report). Ottawa.

Joel Lexchin M.Sc., MD is Professor Emeritus in the Faculty of Health at York University where he taught for 15 years from 2001 to 2016. He is also an Associate Professor in the Faculty of Medicine at the University of Toronto, a practising emergency physician in the University Health Network in Toronto and an Affiliate at the University of Sydney. He is a Fellow of the Canadian Academy of Health Sciences. He is the author or co-author of over 200 peer reviewed publications on multiple aspects of pharmaceutical policy. He has been an advisor to the Governments of Ontario, Canada and New Zealand as well as the World Health Organization. He is an editor on two journals. His most recent books are Private Profits versus Public Policy: The Pharmaceutical Industry and the Canadian State (University of Toronto Press, 2016) and Doctors in Denial: Why Big Pharma and the Canadian Medical Profession Are Too Close For Comfort (Lorimer, 2017).

Chapter 6
Professional Ethics, Professionalization, and Regulation of Pharmaceutical Sales Representatives: Analyzing the Costa Rican Case

Gabriela Arguedas-Ramírez

Abstract The purpose of this essay is to present an argument in favor of a strong state regulation of sales representatives working for pharmaceutical companies, taking as an example the Costa Rican legislation, which can be used as part of a strategy to discourage the various incorrect practices that have created a climate prone to corruption in the relationship between prescribers and pharmaceutical companies. This regulation should include three basic mechanisms: (1) The professionalization of sales representatives as a legal requirement, (2) Affiliation to the pharmaceutical professional association in the country, as a mandatory requirement, (3) Creation of state laws or policies to regulate the conduct and practices of pharmaceutical sales representatives, including the necessary mechanisms for filing complaints about breach of such regulations and adequate penalties for both the sales rep and the company. These rules can go along with by transparency policies such as the Open Payments Act in the US, and international guidelines on the proper relationship between prescribers and pharmaceutical companies, such as those published by the WHO.

Keywords Professional ethics · Pharmaceutical · Sales · Costa Rica

6.1 Introduction

The purpose of this chapter is to present an argument in favor of a strong state regulation of sales representatives working for pharmaceutical companies, taking as an example the Costa Rican legislation, which can be used as part of a strategy to discourage the various incorrect practices that have created a climate prone to corruption in the relationship between prescribers and pharmaceutical companies. This regulation should include three basic mechanisms: (1) The professionalization of sales representatives as a legal requirement, (2) Affiliation to the pharmaceutical professional association in the country, as a mandatory requirement, (3) Create state laws or

G. Arguedas-Ramírez (✉)
School of Philosophy University of Costa Rica, San Jose, Costa Rica
e-mail: arguedas.gabriela@gmail.com; maria.arguedasramirez@ucr.ac.cr

K. T. Çalıyurt (ed.), *Integrity, Transparency and Corruption in Healthcare & Research on Health, Volume I*, Accounting, Finance, Sustainability, Governance & Fraud: Theory and Application, https://doi.org/10.1007/978-981-15-1424-1_6

policies to regulate the conduct and practices of pharmaceutical sales representatives, including the necessary mechanisms for filing complaints about breach of such regulations and adequate penalties for both the sales rep and the company. These rules are complemented by transparency policies such as the *Open Payments Act* in the US, and international guidelines on the proper relationship between prescribers and pharmaceutical companies, such as those published by the WHO. A rigorous ethical analysis of corruption in the context of the relationship between prescribers and Big Pharma requires a broader understanding of the effects that neoliberal policies have on health systems and pharmaceutical corporations. Deregulation in certain areas of the pharmaceutical market (control of marketing and drug prices) is correlated with excessive regulation in other areas, such as intellectual property rules linked to free trade agreements, designed to protect the interests and the business model of these companies. Both aspects favor the safeties of the large pharmaceutical industry, to the detriment of the common good, especially the most vulnerable patients.

6.2 Marketing, Big Pharma, and Prescriptions: Overview

The market of medicines depends on the prescriptions that doctors, dentists, and nurses write for their patients. The over-the-counter drug segment only covers a small part of the total drug-therapy offer and, in addition, over-the-counter drugs are priced much less than prescription drugs. Therefore, pharmaceutical companies need to maximize the number of prescriptions and the price of medicines to obtain increasing profits. This business model is justified by Big Pharma on the premise that there is no other way to finance the research and development of new medicines.

Several scholars in the history of science and medicine have placed the marketing of medicines within the context of the broader process of industrialization, standardization and increasing dependence of pharmacy on laboratory techniques. Gaudillière and Thoms (2015) analyze the specialized literature that explains the reorganization of drug manufacturing in the post-war period in Europe. First, there is a change of scale, facilitated by an expanding market thanks to the expansion of health insurance. Then, this market impulse facilitates mergers between small family industries, founded by pharmaceutical professionals. Another characteristic of this transformation is the entry into the market of new classes of drugs, which were effective in diseases for which no effective treatment had been achieved. In addition, there is a change in regulation, becoming more important the administrative regulation than the professional.

Big pharmaceutical companies used to have a remarkable influence on basic research for the development of drugs. Nevertheless, that has changed as more actors are involved in these processes, such as universities and research centers financed with public funds. However, the clinical trials needed to prove efficacy and safety still depend, in large part, on the funding that comes from these powerful companies. Because of the big pharmaceutical companies' lobby, international regulations regarding the ethics of biomedical research, intellectual property and international trade have been shaped, at least, to some degree. For example, in bilateral trade agreements, as CAFTA or NAFTA, the rules applying to pharmaceutical patents rights are tougher than the rules established by the WTO. Those are known as ADPIC-plus rules.

Some of the most widely analyzed and discussed ethical issues vis-à-vis the role of Big Pharma in clinical research are the camouflage of negative results, the concealment of conflicts of interest in the sponsorship of these trials, the design of research protocols that, in order to lower costs, disrespect the rights of participants, and the disproportionate interest in developing new molecules for diseases that have a promising market, leaving aside pathologies that affect more vulnerable sectors of society.

The apparatus for the production and publication of scientific articles that report the results of clinical trials, on which Big Pharma marketing depends, has also been strongly questioned. Scientific journal editors have seen the need to develop and implement rigorous regulations to counteract the influence of corporate interests of pharmaceutical companies on scientific publications. On the other hand, it has been assumed that leaving the prescription as an exclusive medical act is an effective mechanism to ensure the rational use of medications and the protection of patients' health. However, this prescribing system—in the context of a market model applied to health services and technologies—has also caused the interest of pharmaceutical companies in maintaining direct communication with those who determine the sale of their products.

The relationship between pharmaceutical companies and prescribers is complex and has been pointed out as ethically problematic. Some the authors consider that any contact between the companies and the prescribers is itself a source of a conflict of interest and they propose that the mere fact of receiving the visit of the sales representatives of these companies is unacceptable, either because of the risk of influence the habit of prescription to the detriment of the interests of the patient or, simply by investing time in these contacts, which could be used to develop tasks of greater importance and urgency (Gagnon 2013).

This line of argument has supported the implementation of institutional policies that restrict the access of pharmaceutical sales reps to hospitals and clinics and prohibit doctors and medical students from meeting with pharmaceutical sales reps inside the hospital buildings. In the USA, in addition to these preventive measures, a transparency policy called *Open Payments Act* (Sunshine Act) was established as part of the Affordable Care Act, requiring all prescribers to make public the gifts they receive from pharmaceutical companies. Pharma sales reps also seek to meet with health care students, to create a long-term marketing relationship. For some authors, this strategy ensures the progressive normalization of the link between prescribers and pharmaceutical companies, making it more difficult for prescribers to critically analyze the effects of marketing strategies on medical decisions. In this way, the prescription habits of health professionals, which should be based on the ethics of the rational use of drugs, ends up being shaped by the marketing of pharmaceutical companies.

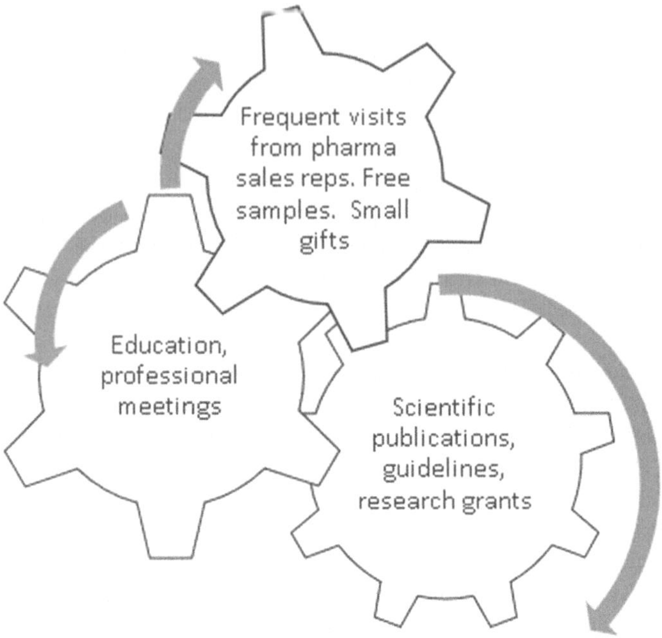

Fig. 6.1 The link between promotional techniques

There are other ethically questionable marketing practices that have become part of the culture of the pharmaceutical business, such as continuing education activities for prescribers. It has been detected that in these activities, payments to lecturers, and the culture of camaraderie and joviality deepen social links between the doctors and the company, which are then used by the company to demand—in very subtle ways—reciprocity to prescribers (Gagnon & Lexchin, 2008; Parker & Pettijohn, 2006) (Fig. 6.1).

6.3 Marketing and Production of Prescription Habits

The various marketing strategies of pharmaceutical companies cover the entire spectrum of techno-scientific activities that make up the drug production chain. As well explained by Petryna et al. (2006), in the prologue of Global Pharmaceuticals, the production of pharmacotherapeutic knowledge is imbued in the primary objective of these companies, which is profit. If the activity of researching, producing and selling medicines is not profitable, at the levels expected by the stakeholders, then that type of business would disappear, regardless of whether there is a social need to purchase medicines.

This explains why there are orphan diseases, orphan drugs, scandals in clinical research, the persistent use of ghostwriters, among other ethically problematic or unacceptable situations. Pharmaceutical companies are organizations based on the principle of maximizing profits and minimizing costs, without this being reflected in the production of better, cheaper and accessible drugs to the population (Sismondo, 2006).

In this political–economic context in which healthcare systems are mostly private (with few exceptions such as UK, France, Canada, and Costa Rica) the various marketing strategies are directed to both the prescriber and the employer. These strategies operate on several levels, from the micro (prescriber) to the macro (boards of directors, general managers, executive directors, purchasing managers, etc.) and they are linked to each part of the conceptualization process and production of the drug, from its experimental phase to its commercialization. ProPublica found that, on average, doctors who receive payments from the pharmaceutical industry prescribe medications differently than those who do not. And the more money they receive, the more brand-name drugs they tend to prescribe. Even those who simply obtained meals from companies prescribed more brand-name drugs, on average. In addition, as payments increased, brand prescription rates also tended to be higher. Physicians who received more than $5000 from companies in 2014 generally had the highest percentages of brand prescription. Among internists who did not receive payments, for example, the average brand prescription rate was around 20%, compared to about 30% of those who received more than $5000. ProPublica analyzed the prescription patterns of physicians, who wrote at least 1000 prescriptions in the Medicare drug program, known as Part D. In five common specialties, as doctors received more money from the drug and device companies, they tended to prescribe a higher percentage of brand drugs (Grochowski-Jones & Ornstein, 2016).

Similarly, recent studies have shown that the influence of Big Pharma also extends to the professional associations that develop the therapeutic guidelines, which are then followed by thousands of professionals in the world, since they are supposed to be a scientific tool for the rational use of drugs (Mitchell, Winn, & Dusetzina, 2018) argue that the source of the ethical conflict in the relationship between Big Pharma and prescribers is the payment to doctors, especially in the development of clinical investigations; and also, in the dynamics with the university hospitals that receive millions of dollars from Big Pharma. Therefore, it cannot be said that the most important problem is contact with pharma sales reps. That would be mistaking the part for the whole.

6.4 The Complex Relationship Between Sales Representatives and Prescribers

Within the conglomerate of marketing tactics, which are applied in almost all countries where there is a private market of medicines dependent on the medical prescription, the pharmaceutical sales representative constitutes the direct link between prescribers and pharmaceutical companies. This link has been central in the framing of allegations of corruption and influence-peddling (Transparency International. Pharmaceuticals and Healthcare Programme, 2016).

The main function of pharma sales representatives is to induce the prescriber (doctor, dentist, nurse in some cases) to develop what in the pharmaceutical corporate world jargon is known as the "prescription habit". They have to build close relationships with the prescribers, based on the technical information that they explain in each visit, and even more important, they are supposed to cultivate that relationship through the delivery of medical samples and certain gifts and reminders of the medication's brand they are promoting.

In addition, pharma sales reps carry out other functions that complement the periodic visits to the prescribers assigned to them, according to a territorial distribution. For example, they participate in continuing education activities that generally include, in addition to the keynote speaker's lecture, an entertainment event, such as concerts, dinners, and shows. That is, continuing education activities (ranging from a one-night activity, to a multi-day symposium or an international congress of a prestigious association) always include an element of fun and socialization. Pharma sales representatives are key agents in the construction of commercial relationships that, in the end, are determinant for sales of prescription drugs. Although in many countries there is publicity and direct to consumer marketing, the promotion of medicines, carefully designed for health care professionals, continues to be a central part of Big Pharma's commercial structure, because without prescriptions their business would not function.

These representatives focus their work in one-on-one contact with health care professionals. Pharma sales representatives usually don't have a degree in healthcare sciences, so they need some training which is provided by the pharmaceutical companies. That way they learn the technical information about the medicines they will promote with the doctors. According to some recent studies, physicians complain that the information these representatives offer is repetitive and sometimes biased, because it omits references to adverse effects or other molecules that may be more effective. Pharma sales representatives who do not have any training in health care sciences, and more specifically, in pharmacology, lack the minimum tools necessary to contrast the information they receive during training and detect inaccuracies or incorrect assertions. Therefore, in communicating with prescribers, they may not be aware of possible omissions or biases. The usual is that pharma sales reps are repeating information that they have memorized and do not fully understand (Norris, Herxheimer, Lexchin, & Mansfield, 2005).

In his ethnographic study in Argentine, Lakoff (2006) explains in detail that pharma sales reps must investigate the prescription habits of the physicians they visit, at least, once a month. To know which medicines doctors are prescribing, sales reps use the databases of companies such as QuintilesIMS, which buy that information in private pharmacies.

That *high contact* described by Lakoff (2006) produces a complex link between prescribers and sales reps, which can even mix affective aspects within this equation. The production of this "habit of prescription" is an intricate process, which goes beyond a mere exchange of information on medicines and cannot be described as just a "purchase of consciences" through lunches or medical samples. Hence the need to introduce greater scientific and professional rigor in this interaction. According to a systematic review published in 2017, of the 19 studies, 15 found a consistent association between interactions that promoted a drug and inadequate increase in prescription rates, lower prescription quality and/or higher prescription costs (Brax et al., 2017). Spending on this form of promotion (detailing) represents a large proportion of the marketing budget of pharmaceutical companies, even though a growing number of doctors in the US and other countries are limiting contact with Big Pharma sales forces (Liu, 2007). Pharmaceutical companies have tried to compensate for this trend of closed doors, using a new strategy called Medical Science Liaison, which gives the appearance of being a more objective and technical link compared to visits by pharma sales representatives.

Abundant literature published since the 1980s[1] demonstrates the influence of pharmaceutical marketing on prescription's patterns, has gradually produced enough political pressure to motivate legislative changes, such as the Sunshine Act (Open Payments Act). However, Larkin et al. (2017) found out that in the USA, the policies that limited access to pharma sales representatives have had a modest effect on prescribing habits in some of the centers that implemented these policies. The decrease in pharma sales representatives is due in part to a decrease in highly successful drugs, but also because a considerably smaller number of doctors are willing to see them. According to a report by ZS, which provides information on access to doctors to drug representatives, 34% of doctors' offices were accessible to drug representatives in 2016, compared to 65% in 2012, before it was created the website (Page, 2017).

The regulation of the promotion and marketing of medicines differs between countries. In the USA pharma sales reps are regulated by the Food and Drug Administration (FDA) and are subject to "fair balance" provisions that require damage, as well as benefit information on all promotion components. Pharma sales reps cannot promote unapproved indications, but they can provide reprints on these uses at the request of physicians. In 2004, France introduced an eight-page pharma sales representatives' letter, which prohibits drug samples, food, gifts and invitations to participate in the studies. Pharma sales reps must provide physicians with approved

[1] See, for example: Physician Motivations for Nonscientific Drug Prescribing, Rebecca Schwartz et al., 1989, Society Medicine, Vol. 28. No. 6, pp. 577–582; Scientific versus Commercial Sources of Influence on the Prescribing Behavior of Physicians, Avorn, Chen and Hartley, Medicine, Science and Society, 1982, Vol. 73.

product information. In Canada, the regulation of pharma sales reps is largely delegated to Canada's Research-Based Pharmaceutical Companies (Rx&D). The Rx&D code of ethics requires consistency with the approved product information and current medical thinking. The federal regulatory agency, Health Canada, can exercise its legislative authority if necessary, but rarely does it in practice (Mintzes, Lexchin, Sutherlang, Beaulie, & Wilkes, 2013) However, this study showed that minimally adequate safety information "did not differ in the US and Canadian sites, despite regulatory differences."

6.5 The Regulation in Costa Rica

In Costa Rica, anyone who wants to work as a pharmaceutical sales rep, needs a general degree in Pharmacy or Medicine. Pharmaceutical sales representation is part of the pharmaceutical or medical professional act, because, according to the *Ley General de Salud* (the Health Care national legislation), only professionals in these two areas can exercise that occupation. Pharma sales reps must be duly registered in the Costa Rican College of Pharmacy or the Costa Rican College of Physicians. In the veterinary pharmaceutical companies, the requirement is to be a veterinarian. The set of rules regarding the job of pharmaceutical sales representatives include not just the general healthcare legislation, but also the specific regulation established for all registered pharmacists or physicians, according to their codes of professional ethics and any other regulations that apply for those professions. On the other hand, Costa Rica lacks a piece of legislation to regulate certain commercial and marketing practices of the pharmaceutical industries, as there is in the US and in other countries. There is no legal requirement for doctors to make public the data on the gifts and incentives they receive from pharmaceutical companies. This is a normative weakness in the control and prevention of corruption within the relationship between prescribers and companies.

However, for the purposes of this chapter, I will concentrate on analyzing the particular way in which the work of the pharmaceutical sales reps is regulated in Costa Rica and the lessons that can be drawn from this case, to enrich the global discussion on the risk of corruption in the relationship between pharmaceutical companies and prescribers.

6.6 Sales and Marketing as Part of the Pharmaceutical Professional Practice

Analyzing the regulation of pharmaceutical sales reps in Costa Rica reveals certain peculiarities that are worthwhile to consider, in the general discussion about the ethical conflicts between pharma companies and prescribers.

The legal requirements to work as a pharma sales rep in Costa Rica was established in 1974. Prior to that moment, anyone with a higher education degree could work as a pharma sales rep, as is still the case in almost all countries where pharmaceutical companies have this type of salesforce. Costa Rica is one of the few countries in the world in which a law regulating health services establishes the legal terms to work as a pharma sales rep. In most countries in the Americas, pharma sales reps do not need an academic degree in the healthcare sciences to get the job. Pharmaceutical companies oversee giving them the necessary training to promote the medicines to the prescribers and the pharmacists. These trainings focus on the development of two set of skill: (1) the ability to memorize the minimum scientific information necessary to speak with apparent confidence, about the medication they are promoting (2) the ability to persuade the prescriber to use that medication in their patients.

Pharmaceutical companies (whether they sell innovative drugs or generic ones) give more emphasis to the process of training their sales forces in the second group of skills. Pharma sales representatives are not properly vendors of pharmaceutical products. Their business is similar to political lobbying or public relations. Its main purpose is to convince the doctor or the dentist (or in some cases, the nurse) to prescribe the medication and to adopt it as part of the therapeutic arsenal of their confidence.

In Costa Rica, the institutional and academic narrative about *visita médica* does not coincide with this explanation about the role of the pharma sales rep. Pharmacy Schools in Costa Rica have contributed to build up an idea about the implementation of the *visita médica* as a technical-scientific activity, more in line with the professional profile of the pharmaceutical and medical field. Pharmacy students are told that the *visita médica* (in Spanish the term does not correspond at all with the representative sales, in English) is a professional act focused on scientific divulgation and advice on the rational use of medicines. In this way, the pharmaceutical professional ethos also includes the field of action of the pharma sales rep. This does not mean that the pharma sales reps in Costa Rica do not respond to the pressure of each pharmaceutical company, to increase the sales of their medicines. The difference with countries where there is no regulation like this lies in the existence of a unifying element that has favored the creation of a culture of *visita médica* that, to some extent, resists the mere commercial logic of sales reps in the US and other countries. In the debates of the pharmaceutical community, the obligation to ensure that the medical visit remains an ethical and technical activity is insisted on. ASVIMED's president (Association of Medical Visitors of Costa Rica, is a professional organization that gathers all *visitadores medicos*, the majority of whom are professionals in Pharmacy, but there are also some doctors), indicates that

> I have always believed that, unfortunately, the pharmaceutical market in general faces the greatest ethical dilemma of all health professions. This is the sector in which economic interests are most strongly opposed to the interests of patients. And of course, the practice of the pharma sales reps does not escape this. So, in summary, the main ethical dilemma is to sell the best therapeutic alternative for the patient, but at the same time meet the demands and goals of profitability of the companies. (private communication)

Pharmacy schools in Costa Rica are integrating courses on *visita médica* and ethical marketing. The incorporation of these classes confirms that the *visita médica* is associated with a specialized professional practice, which must be kept in line with the scientific and ethical standards of the pharmaceutical field. In turn, this has effects on the construction of an ethos and a professional identity. By working as pharma sales representatives, they are not exercising just a marketing function alone, but also—and more importantly—they are practicing as Pharmacy professionals. Of course, in no way does the association between the pharmaceutical profession and the *visita médica* constitute an infallible mechanism against corruption relation between pharmaceutical companies and prescribers. But it is indisputable that to achieve the highest professional level of communication between prescribers and sales reps, academic training is a determining factor.

Some of the most experienced *visitadores médicos* in the country with whom I have had several conversations about this topic reported that one of the most pressing problems at this moment is the growing number of representatives, that has been increasing considerably in recent years, putting much more pressure on both the *visitadores médicos* and on the prescribers. This increase—according to the opinion of some them—has emboldened ethically questionable practices, such gifts giving and other kinds of monetary incentives for prescribers who accept to meet with them. It seems that to obtain the privilege of being received by prescribers, companies are willing to do whatever is not considered illegal in the country, even though it is not legal in the country where their headquarters are. Besides that, this situation gets more complicated due to the disparate quality of academic training that exists between universities (in Costa Rica, public universities have more prestige and are better ranked than most private universities, but is getting more difficult for students to be admitted at these public universities and just one of them has a Pharmacy School.

Costa Rican regulation involves controlling the weakest part of the chain of influence over prescribers: pharma sales reps. At higher hierarchical and institutional levels, there are other ways to coerce or influence prescribers or drug buyers. What is discussed in this article is the possibility of including stricter controls over pharmaceutical sales representatives, within the range of state norms and policies that should sanction or prevent corruption (Table 6.1).[2]

[2]Regarding the strategy of pharmaceutical companies to maximize their sales to the State, by using the legal system to force the purchase of medicines that are not included in the official list of medicines of the public health system, I recommend reading Angelina Godoy's book: Of Medicines and Markets, published by Stanford University Press in 2013.

Table 6.1 Summary of the legal framework that regulates the medical visit in Costa Rica

Law/Decree	Name	What is regulated
Ley No. 5395	National Health Care Law	ARTICLE 140. The sale and trade of free samples and their possession in pharmacies, health kits, or retail stores is forbidden. Free samples and advertising material for medicines must be delivered only to professionals in the health care sciences, by duly accredited pharmaceutical sales representatives (visitadores médicos) who must be members of the College of Physicians and Surgeons or of Pharmacists. Likewise, regarding medicines for veterinary use the sales representative must be a member of the College of Veterinary Doctors or the College of Pharmacists. The supplied information must contain at least the complete list of active ingredients, their proper form of administration and their counter indications
Executive Decree No. 26374-S	Pharmaceutical sales representatives regulation	ARTICLE 6 In the exercise of their professional activity, the medical visitor must comply with the following standards: (a) Do not promote the commercialization of free medical samples (b) Deliver free medical samples only to professionals in the Health Care Sciences, with the sole purpose of supporting the information on the product (c) Conduct in accordance with the high professional image of a member of the College of Pharmacists, Physicians and Surgeons and Veterinary Physicians of Costa Rica (d) Keep the cordiality and respect, characteristic of their professional work, with patients and staff of the clinic or hospital they visit (e) Refrain from using negative propaganda against its competitors, to promote its products (f) Not offer gifts, gratuities or compensation, in exchange for the prescription of the product (g) Safeguard promotional material, to ensure that it cannot be stolen or misused (h) Refrain from disclosing other information than the one authorized in this regulation ARTICLE 9 The Ethics Committee of the respective College will determine any transgression of this regulation

(continued)

Table 6.1 (continued)

Law/Decree	Name	What is regulated
Executive Decree No. 4917 (1975)	Import and use of free medical samples	
Regulation based on the Law of the National College of Pharmacy, approved by the General Assembly on 2015	Professional Code of Pharmaceutical Ethics	ARTICLE 68. Pharmacy professionals who work in the field of pharmaceutical sales must have the highest ethical behavior and adequate scientific and technical training, to carry out the activities of promotion of pharmaceutical products in a correct and responsible way, to contribute to the protection of public health. They must always keep equanimity and a good presentation, in accordance with the high professional image of a member of the Costa Rican College of Pharmacy ARTICLE 69. Pharmacy professionals working as pharmaceutical sales representatives must maintain a constant updating and professional improvement, to provide a better service to professionals in the health care sciences (…) ARTICLE 70. Pharmacy professionals working as pharmaceutical sales representatives shall only deliver medical samples to professionals in the health care sciences with the exclusive purpose of supporting the information on the product, for which reason they should not carry out the direct sale of the product nor allow or promote the commercialization of the product ARTICLE 71. Pharmacy professionals working as pharmaceutical sales representatives will not encourage or permit unauthorized persons to perform the functions of pharma sales reps in any way. In case of knowing about any unauthorized activity of the sort, they must make the pertinent complaints at the College of Pharmacy or at the judicial authority, for illegal exercise of the profession. All pharmacists are obliged to report a pharmaceutical sales representative who is not authorized by law ARTICLE 72. Pharmacy professionals working as pharmaceutical sales representatives cannot offer or promise gifts, monetary or in-kind benefits, money in cash, trips or lodgings to health care professionals or support staff, to encourage the prescription and sale of medicines. The educational material must comply with the established legal requirements ARTICLE 73. Pharmacy professionals working as pharmaceutical sales representatives cannot promote drugs for indications not approved by the Ministry of Health ARTICLE 74. The professional person in pharmacy can only provide scientific information, real and objective

(continued)

Table 6.1 (continued)

Law/Decree	Name	What is regulated
Directive No. 31407-5-A-09. Signed by the Medical Direction of the National Public Health Care System (CCSS) on 2009	Directive for all public health care centers	Health care professionals, as public officials, must act as such in the institutional sphere. The CCSS must ensure transparency, and for this reason, they must refrain from receiving representatives of pharmaceutical companies during the working hours of the institution. Therefore, visits by representatives of pharmaceutical laboratories and representatives of drug distributors to the facilities of the CCSS are forbidden. Representatives of pharmaceutical laboratories and representatives of drug distributors are must not promote, market or distribute pharmaceutical products to users within the facilities of Health Care Centers
Legal Department. CCSS Official communication DJ-079-2013. (2013)		(…) Because the health care services of the CCSS are public and are regulated by the principles of continuity and efficiency, stated in article 4 of the General Law of Public Administration, therefore, everything that is not expressly authorized is forbidden to public officials. Consequently, it should be noted that physicians are hired by the Institution to carry out the activities related to their profession, which do not include attending to pharmaceutical sales representatives during working hours

6.7 Discussion: The Suggested Model; State Regulation, Professionalization, Union Regulation

Paul Hunt, as Special Rapporteur on the right to health, in his report A/63/263, explains that:

> Ministers, senior public officials and others have argued that the policies and practices of some pharmaceutical companies constitute obstacles to States' implementation of the right to the highest attainable standard of health and, in particular, their endeavors to enhance access to medicines. They have mentioned, for example, excessively high prices, inadequate attention to research and development concerning diseases that disproportionately impact people in developing countries, inappropriate drug promotion, and problematic clinical trials. Ministers and senior public officials have also acknowledged, however, that the pharmaceutical sector has an indispensable role to play in relation to the right to health and access to medicines. Moreover, they have recognized the constructive contribution of specific pharmaceutical companies. (Hunt, 2008, 7)

Since 1988, the WHO Assembly adopted the document Criteria for Medicinal Drug Promotion that establishes ethical principles and suggested technical criteria for member countries to regulate the marketing activities of pharmaceutical companies, including the use of medical samples, activities of continuing education and the activities of the pharma sales reps.

The guidelines have already emphasized, in general terms, the central importance of transparency in relation to access to medicines (Guidelines 6–8). Guidelines 39–41 apply this general principle of transparency to the specific context of ethical marketing and promotion. Promotion and commercialization give rise to a wide range of problems of access to medicines, such as advertising for health professionals and the general public, packaging and labeling, and information for patients. Based on ethical considerations, the World Health Organization's Criteria for the Promotion of Medicines provide authoritative guidance on these important issues (Guideline 40). p. 24

These regulation criteria refer to

1. *The minimum training and basic knowledge in medical issues that the reps should have, as well as the ethical orientation necessary to perform their work with integrity.*
2. *The duty of sales reps to offer information to prescribers that is complete and truthful, free of brains and based on scientific evidence.*
3. *The duty of the employers to assume responsibility for the actions of the sales reps that work for them.*
4. *The obligation of sales reps not to offer any kind of prize or encouragement to prescribers and dispensers, to use the medicines they promote. To avoid bad practices, it is recommended that most of the salary of sales reps do not come from sales volume.*
5. *The duty of the companies to publicly disclose their promotion and commercialization policies and activities, including costs.*

From a human rights' perspective and, specifically, taking the human right to health as a starting point, the promotion of medicines must address the particularities of these goods, which are essential to preserving the life and health of people. In addition, doctors must abide by the principles of medical ethics to make decisions about the prescription for their patients. The prescribers have duties towards their patients, which include the duty to make therapeutic decisions that benefit each patient, according to their needs and situations. By virtue of this duty, prescribers have the obligation not to allow commercial strategies to alter the decision-making for a rational and individualized use of medicines.

Medical organizations, such as *Médicos sin Marca*[3] or *No Free Lunch*[4] advocate prescribing by the generic name of the drug, and not by brand. This can be a solution to avoid the influence of pharmaceutical companies in the habit of prescription, but this is not enough to limit the great power of influence that pharmaceutical corporations have over the whole process of production and sales of medicines.

Gagnon and Volesky (2017) warn that we should not naively trust that the solution lies in the generic market. *Médicos sin Marca* and other similar organizations consider that the main cause of the problem is the use of brand names, so they prescribe and promote the prescription by generic name. This, however, does not solve the growing concentration of the market in a few companies, some of them players in both the market of innovators and the generic market. Besides, these companies have other strategies to ensure their sales, through the negotiating conditions with the distributors and pharmacies (drugstores).

Proposing a regulation of pharmaceutical marketing and the functions of sales reps, that could contribute to eradicating corruption in the context of promoting drugs to prescribers, demands the de-normalization of the political economy of medicines' production, which is embedded in neoliberal capitalism. Several authors have referred to some of the variables of the political economy of drug production, such as financial incentives and the tailored legal framework.

Nonetheless, the contrast between how the concept of medicine is understood in the capitalist economic context, and the way a therapeutic drug is conceptualized in the field of public health, has not been placed at the center of the analysis. These two diametrically opposed forms of conceiving the same object need to enter the debate and should be made explicit in the discussion about corruption in the global drug market.

From a public health and human rights perspective, drugs are necessary technologies to respond to the need of someone who is ill. From the free market's logic, drugs are consumer goods, whose production and sales follow the rules of supply and demand. And those who produce that good, expect not only an economic gain, but—according to the corporate narrative, the maximum profit is expected, in the shortest time possible, with the minimum investment necessary.

Clearly, this contrast produces a conflict between the legitimate interests of a person who needs to restore her health and a company that follows the rules of the

[3]http://www.nogracias.eu/2012/06/04/medicos-sin-marca/.

[4]http://www.nofreelunch.org/.

contemporary capitalist economic game. In addition, medicines do not follow a flow of exchange similar to the rest of goods and services that are traded in the market. The patient does not decide which medicine to buy, who decides that is the prescriber (doctor, or other specialist), who is supposed to make that decision not based on the price of the medication, but on the efficacy and safety features that best respond to the particular situation of the patient: this is what is known in the pharmaceutical and public health field as the rational use of medicines (Groves, Sketris, & Tett, 2003).

As Gagnon (2013) has well explained, corruption is institutional:

> It is of little use to blame the pharmaceutical companies for lacking corporate social responsibility if current financial incentives promote such practices. In fact, because these incentives are systemic, the concept of institutional corruption can be used to describe the dynamics of influence at work in the drug industry, leading the institutions underpinning medical research and physicians' prescribing behavior away from their ethical purposes, even when quid pro quo corruption itself is not involved. To tackle this problem, then, one must change the systemic financial incentives at work. (Gagnon, 2013)

Further complicating the analysis, Feldman, Gauthier and Schuler (2013) have proposed the thesis that there are two separate communities within the pharmaceutical industry: researchers and managers:

> We argue that the current approach to pharmaceutical monitoring is designed mainly to deal with the rational-choice misconduct of executives rather than with the misconduct of researchers, of which the researchers themselves are often only partly aware, due to blind spots which we discuss below. We document some differences between misconduct by executives and researchers and show how concepts from behavioral ethics — explaining how good people unknowingly engage in bad behavior — could explain much of the misconduct carried out by pharmaceutical researchers. We conclude with some normative ideas of how to regulate pharmaceutical researchers as a community. (Feldman, Gauthier, & Schuler, 2013), 621

I agree with that claim and add that, if the work of the pharma sales reps is professionalized, there would be a third community, whose regulation is vital to avoid an inappropriate interference in the medical act of prescription that could go against the rational use of medicines. If pharma sales reps are trained pharmacists, associated with a professional board, they must adhere to the rules of conduct established for that organization.

Based on the critique of the political economy of the production of medicines, as mentioned before, I consider that the intervention of the State is indispensable, as a minimum requirement for an effective public policy that de-normalize and eradicates the influence of pharmaceutical corporations in the act of prescribing medicines. This intervention would not be limited to the issuing of regulations, but rather implies some minimal degree of control and sanctions' application.

Empirical evidence has shown that self-regulation is not effective. Self-regulation, through the Pharmaceutical Marketing Code of Practice of the International Federation of Pharmaceutical Manufacturers and Associations (IFPMA), complemented by the member association and the Company's codes, is the industry's response to ensure that the standards are adequately met in this respect. However, research in both Europe and emerging markets shows that this system is not protecting consumers and healthcare professionals from biased and misleading information from pharmaceutical companies (Bala-Miller, Macmullan, & Upchurch, 2007).

Habibi et al. (2016) found out in their study that, unlike industry respondents, regulators did not see sales visits as an essential source of information for prescribers and recognized their potential for misinformation and irrational prescribing. Despite these reservations, they also believed that sales visits were within the scope of permitted marketing practices. Taking this evidence into account, the proposed regulation could help improve the quality of a sales representatives' message to health care professionals, making it informative, balanced, and useful. According to one of the health professionals, interviewed for this study, there is a concern about professional ethics being compromised: "I think our patients would be horrified if they knew to what extent some doctors receive … incentives [dinners, trips and tickets] of pharmaceutical companies." This study determines that both health professionals and consumer advocates doubt the authenticity and effectiveness of current regulations (Habibi et al., 2016).

In 2002, a study conducted for WHO arrived at similar conclusions: sanctions are weak or nonexistent, and there is no accessible and reliable information about the sanctions applied, in those countries where they apply:

> If active monitoring is carried out only sporadically, with minimal in-depth examination of drug advertisements and promotion, it is unlikely to be effective. If there are no sanctions, or only small fines are imposed when a violation is discovered, then the deterrent effect is minimal. For example, a pharmaceutical company that violates the law may have to pay nothing more than a small fine. In that case, it may be more cost-effective from the company's point of view, given the large amount it has already spent on advertising, to pay the fine for an extended period of time rather than withdraw the advertisement. (Ratanawijistrasin & Wondemagegnehu, 2002), p. 105

As explained before, unlike countries such as the USA and Canada, in Costa Rica there is are legal and ethical directives that establish the limits of a legitimate implementation of pharma sales reps. This legitimacy is, in turn, linked to the ethical and legal parameters of the pharmaceutical profession and the medical profession. It would be a mistake to assume that such regulation is enough to control the pressure of pharmaceutical companies on prescribers and patients; however, it does constitute a restraint that could be effective, if used by social and state actors.

The creation of a culture of transparency and accountability, in the professional practice of medicine and pharmacy, starts from the stage of university education. If, as part of this training, the component of the *visita médica* (pharmaceutical sales) is included, there is an opportunity to permeate and forge an ethics of the pharmaceutical sales representation that adheres to pharmaceutical professional ethics. That is not possible to do if any person can exercise this kind of job. Nor is it possible to rely on the training that companies offer to their sales forces. Corporate pressure is a factor that will affect the pursue of a strong culture of transparency and accountability, but as a counterweight, professional associations and public health institutions can impose controls, limits, and supervision.

References

Bala-Miller, P., Macmullan, J., & Upchurch, L. (2007). *Drugs, doctors and dinners. How companies influence health in the developing world.* London: Consumers International.

Brax, H., Fadlallah, R., Al-Khaled, L., Nas, H., El Jardali, F., & Akl, E. (2017). Association between physicians' interaction with pharmaceutical companies and their clinical practices: A systematic review and meta-analysis. *PLoS ONE, 12*(4). Retrieved from https://journals.plos.org/plosone/article?id=10.1371/journal.pone.0175493.

Feldman, Y., Gauthier, R., & Schuler, T. (2013). Curbing misconduct in the pharmaceutical industry: Insights from behavioral ethics and the behavioral approach to law. *Journal of Law, Medicine and Ethics, 41,* 620–628.

Gagnon, M., & Lexchin, J. (2008). The cost of pushing pills. A new estimate of pharmaceutical promotion. *PLoS Med.* Retrieved from https://www.medscape.com/viewarticle/571493_print.

Gagnon, M.-A. (2013). Corruption of pharmaceutical markets: Addressing the misalignment of financial incentives and public health. *Journal of Law, Medicine and Ethics, 41,* 571–580.

Gagnon, M.-A., & Volesky, K. (2017). Merger mania: Mergers and acquisitions in the generic drug sector from 1995 to 2016. *Globalization and Health, 13,* 62.

Gaudillière, J.-P., & Thoms, U. (2015). *The development of scientific marketing in the twentieth century.* New York: Routledge.

Grochowski-Jones, R., & Ornstein, C. (2016). *Matching industry payments to medicare prescribing patterns: An analysis for propublica.* Retrieved from https://static.propublica.org/projects/d4d/20160317-matching-industry-payments.pdf?22.

Groves, K., Sketris, I., & Tett, S. (2003). Prescription drug samples—Does this marketing strategy counteract policies for quality use of medicines? *Journal of Clinical Pharmacy and Therapeutics, 28,* 259–271.

Habibi, R., Guénette, L., Lexchin, J., Reynolds, E., Wiktorowicz, M., & Mintzes, B. (2016). Regulating information of allowing deception? Pharmaceutical sales visits in Canada, Frances and the United States. *Journal of Law, Medicine and Ethics, 44,* 602–615.

Hunt, P. (2008, August 11). *Interim report of the Special Rapporteur on the right of everyone to the enjoyment of the highest attainable standard of physical and mental health.* Geneva: United Nations General Assembly.

Lakoff, A. (2006). High contact: gifts and surveillance in Argentina. In Global Pharmaceuticals. Ethics, Markets and Practices. Petryna, A., Lakoff, A., & Kleinman, A. (Eds). Duke University Press, 111–135.

Larkin et al. (2017). Association between academic medical center pharmaceutical detailing policies and physician prescribing. *JAMA, 317*(17), 1785–1795. https://jamanetwork.com/journals/jama/fullarticle/2623607.

Liu, Q. (2007, October). The dynamics of competitive drug detailing. *The Johnson School at Cornell University.* Ithaca, USA. Retrieved 2017, from https://pdfs.semanticscholar.org/6055/3b04014857d9d7fdd5bdfa538bd9e5c2dc74.pdf.

Mintzes, B., Lexchin, J., Sutherlang, J., Beaulie, M. D., & Wilkes, M. (2013). Pharmaceutical sales representatives and patient safety: A comparative prospective study of information quality in Canada, Frances and the United States. *Journal of General Internal Medicine, 28,* 1368–1375.

Mitchell, A., Winn, A., & Dusetzina, S. (2018). Pharmaceutical industry payments and oncologists. *JAMA Internal Medicine, 178,* 854.

Norris, P., Herxheimer, A., Lexchin, J., & Mansfield, P. (2005). *Drug promotion: What we know, what we have yet to learn. Reviews of materials in the WHO/HAI database on drug promotion.* London: World Health Organization and Health Action International.

Page, L. (2017, November 21). *Sunshine act: Are doctors still enjoying free lunches from drug reps?* Retrieved from Medscape. https://www.medscape.com/viewarticle/885792_2.

Parker, R. S., & Pettijohn, C. E. (2006). Pharmaceutical drug marketing strategies and tactics. *Health Marketing Quarterly, 22,* 27–43.

Petryna, A., Lakoff, A., & Kleinman, A. (2006). *Global pharmaceuticals. Ethics, markets, practices*. Durham: Duke University Press.
Ratanawijistrasin, S., & Wondemagegnehu, E. (2002). *Effective drug regulation. A multicountry study*. Geneva: WHO.
Sismondo, S. (2006). Ghosts in the machine: Publication planning in the medical sciences. *Social Studies of Science, 39,* 171–198.
Transparency International. Pharmaceuticals and Healthcare Programme. (2016). *Corruption in the pharmaceutical sector. Diagnosing the challenges*. London: Transparency International UK.

Gabriela Arguedas-Ramírez is an Associate Professor of Philosophy and Women's Studies at the Universidad de Costa Rica, where she teaches Bioethics and Feminist Topics. She holds a professional doctorate in Pharmacy, a master's degree in Bioethics and is currently finishing a Ph.D. in Cultural Studies, working on the issues of hunger, justice and subjectivity in Central America. She has published articles about the ethics of genetic patents, obstetric power/violence, ecofeminism, religious fundamentalism and sexual rights. She is the author of Situación de los Derechos Humanos de Poblaciones Históricamente Discriminadas en Costa Rica (Instituto Interamericano de Derechos Humanos, 2013). In 2015, as part of an alliance between the Center for Justice and International Law (CEJIL) and the Women's Studies Research Center (CIEM) at the University of Costa Rica, she presented the results of her research on obstetric violence in Costa Rica, in a thematic audience for the Inter-American Commission on Human Rights. She is currently working with Dr. Steven Miles, professor emeritus at the University of Minnesota, in the translation to Spanish of his book, Doctors who Torture: The Pursue of Justice.

Chapter 7
Exploring Accountability and Transparency Within International Organizations: What Do We Know and What Do We Need to Know?

Sandani Hapuhennedige, Emma Charlotte Bernsen and Jillian Clare Kohler

Abstract Many international organizations strive to maintain good governance within their organizations and in their country programming activities as a means to reduce corruption. When good governance is in place, corruption is less likely to happen given the higher levels of accountability and transparency that can expose corruption and inefficiency. Despite these efforts, the past two decades have witnessed a striking emergence of corruption on a global scale, raising concern across many international organizations. The main objective of this research was to probe inquiry on good governance, explore the current scope of strategies that are known to be most effective, and discern any best practices that can be used to combat the issue of corruption. A literature review was conducted to specifically explore the past and current dialogue on governance, accountability, and transparency (GAT) mechanisms within international organizations (IOs). This study involved a search of three databases (ProQuest, EBSCO, and Social Sciences Citation Index). Since 1971, there has been broad discussion and analysis of GAT strategies within IOs. While no 'best practices' for GAT were found, the selected papers predominantly discussed the gaps in GAT, the erosion of GAT as concepts, and the declining legitimacy and credibility of these efforts. Consequently, this paper documents the changes in conceptualizations and the current state of GAT mechanisms and strategies. This study further reveals the need for a new wave of scholarship and inquiry and introduces the practice of evaluation as a potential opportunity to strengthen good governance. These collated findings probe thinking on crucial next steps for IOs, and the importance of establishing a stronger understanding of what mechanisms work best in practice and where. As we seek to meet the anti-corruption objectives of Sustainable Development Goal #16, building our knowledge in this area is imperative.

Keywords Governance · Good governance · Accountability · Transparency · International organizations · Anti-corruption · Evaluation

---

S. Hapuhennedige · E. C. Bernsen · J. C. Kohler (✉)
University of Toronto, Toronto, Canada
e-mail: jillian.kohler@utoronto.ca

K. T. Çalıyurt (ed.), *Integrity, Transparency and Corruption in Healthcare & Research on Health, Volume I*, Accounting, Finance, Sustainability, Governance & Fraud: Theory and Application, https://doi.org/10.1007/978-981-15-1424-1_7

7.1 Background

The issue of corruption and the development of public policy to mitigate the effects of corruption have gained prominence globally in the past two decades. For example, the United Nations Convention against Corruption (UNCAC), which was adopted by the United Nations General Assembly in October 2003 and came into force in 2005, raised the importance of fighting corruption worldwide. In 2009, the United Nations Secretary-General Ban Ki Moon even highlighted the impact of corruption on the Millennium Development Goals (MDGs). He emphasized that corruption can kill development and may very well impede efforts to achieve the MDGs (UN, 2009). The follow up to the MDGs, the Sustainable Development Goals (SDGs), significantly includes Goal #16.5, which directs attention to reducing corruption and bribery (UN 2015); this is situated within the broader Goal #16 of promoting peaceful and inclusive societies.

Addressing corruption matters as the global community seeks to improve equity. There is no doubt that the poor have a greater reliance on public services and corruption places a disproportionate burden on the poor (Kohler, 2011). Good governance is a critical component of anti-corruption and is a growing priority for international institutions and their member governments alike. During the past two decades, there has been a "corruption eruption," in the form of a striking emergence of corruption on the global agenda (Wang & Rosenau, 2001). This has been driven by the mounting costs that bad governance and corruption bring in an era of globalization to economic development and growth, the financing of terrorism, and trust in the effectiveness and legitimacy of democratic and other governments themselves. Corruption is costly, estimating US $2.6 trillion, which is more than 5% of the global GDP (OECD, 2014). Other estimates suggest that bribes paid alone can equate to over US$ 1.5 trillion each year (World Bank, 2017). Corruption increases the cost of doing business by up to 10% on average and it is estimated that each year 20–40% of development assistance is stolen through high-level corruption from public budgets in developing countries (OECD, 2014). Thus, many international organizations have increasingly tried to control corruption in a variety of ways, such as through integrity offices, oversight offices, general accountability frameworks and broader inclusion of civil society (e.g., transparency) in its operations. Good governance, alongside the principles of accountability and transparency, are potential strategies for combating corruption. Examples include the World Bank, the World Health Organization (WHO), and the United Nations Development Program (UNDP).

7.2 Examples of Good Governance Initiatives

Most United Nations (UN) organizations and international financial institutions have independent integrity offices focused on identifying and addressing fraud and corruption. Since April 2010, the multilateral development banks (e.g., World Bank, the

African Development Bank Group) have had an agreement in place for the enforcement of debarment decisions[1] (IFI, 2010). This agreement harmonizes the standards for applying sanctions for fraud and corruption and mutual recognition amongst the Asian Development Bank, European Bank for Reconstruction and Development, the World Bank, the Inter-American Development Bank, and the African Development Bank. Practices that are liable for sanctions are those that are defined as corrupt, fraudulent, collusive, or coercive. Within the World Bank, corruption issues are addressed by the **Integrity Vice Presidency (INT)**, an independent unit charged with investigating and pursuing sanctions related to allegations of fraud and corruption in Bank Group financed activities. To improve its accountability, the World Bank Group is also seeking to include **beneficiary feedback** in all of its projects by 2018. This is part of a larger Bank strategy to increase citizen participation in its projects through its Strategic Framework for Mainstreaming Citizen Engagement in World Bank Group Operations. The World Bank also established the **Global Partnerships for Social Accountability (GPSA)** in 2012 to promote citizen input in any challenges in service delivery, and to strengthen the performance of public institutions in developing countries. GPSA provides funding as well as information to civil society organizations in its member countries. GPSA has 240 partners and 45-member countries. The Bank also has an Internal Audit Vice Presidency (IAD) which is an independent unit that evaluates the WBG's governance, risk processes, and internal management. IAD publishes Annual Reports which summarize fiscal year results, as well as governance issues.

The WHO is subject to both external and internal audits. For example, The Office of Internal Audit and Oversight (IAO) inspects, monitors, and evaluates the WHO's internal control, financial management, and is also responsible for addressing any alleged breaches. The External Auditor monitors the WHO's operations in terms of its financial risk management and the efficacy of the organization's internal control system. Additionally, there is an independent advisory group that is organized through the Executive Board of the WHO. The purpose of the Independent Expert Oversight Advisory Committee is to advise the Program, Budget and Administration Committee and through it the Executive Board, in fulfilling their oversight advisory responsibility; on request, it also advises the Director-General on issues within its mandate. Last, a relatively recent addition (established in January 2014) to monitoring accountability within the WHO is the Office of Compliance, Risk Management and Ethics (CRE). CRE's cited aims are to promote transparency and accountability through improving compliance, managing a risk framework, and promoting ethical values (WHO, 2015).

The UNDP, along with UNFPA (United Nations Population Fund) and UNOPS (United Nations Office for Project Services) employ the same agreed-upon definition of accountability. As such, accountability is understood as the requirement to: (i) demonstrate that work has been performed in accordance with agreed-upon standards and (ii) fairly report on performance related to "mandated roles and/or

[1]Known as "Agreement for Mutual Enforcement of Debarment Decisions among Multilateral Development Banks"

plans" (UNDP, 2008, p. 4). This common definition, adopted across these agencies, is helpful *insofar as it clearly defines institutional expectations in this area*. UNDP's accountability was established by the General Assembly Resolution 26/88, affirmed by Resolution 29/250, and reaffirmed by Resolution 62/208. The UNDP is held accountable to participating countries; this includes project beneficiaries as well as its donors. To support accountability, the UNDP has an accountability framework and oversight policy. The accountability framework emphasizes UNDP's commitment to results and risk-based performance management, as well as values and culture of accountability and transparency; this is *defined as a process by which reliable, timely information about existing conditions, decisions, and actions relating to the activities of the organization is made accessible, visible and understandable* (UNDP, 2008, p. 5).

UNDP's oversight policy consists of specific procedures, tools and a reporting schedule for providing UNDP management and its stakeholders, independent reviews, including evaluation at all levels of UNDP programmatic interventions. There are independent internal and external oversight processes that verify if systems of control are in place including evaluation of the policy framework, efficient utilization of resources, and adherence to professional and ethical standards (albeit no mention of how the latter standards are measured is provided).

The UNDP notes that it has many processes in place to strengthen its accountability and transparency in addition to the above. These include an integrated financial resource framework; strengthening its resident coordinator system, processes, and tools for enterprise-wide risk management; investing in staff professional programs; regular review of policies and guidelines about roles, responsibility, and accountability; fact-based monitoring mechanisms; an independent audit advisory committee; appointing a Head of the Ethics Office; the promotion of ethical conduct; and addressing allegations of abuse. However, the relative strengths and weaknesses of these processes in achieving the large goals of accountability and transparency are not detailed in its public documents. Last, the UNDP, along with other agencies in the United Nations System, has begun to adopt the International Public Sector Accounting Standards (IPSAS) since 2013. These standards were adopted by the UN General Assembly in 2006 and replaced the United Nations System Accounting Standards.

The Administrator of UNDP is held responsible for overseeing the operations of the UNDP and is accountable to the Executive Board for all management and activities of the UNDP ***including accountability for the associated funds and programs administered by the UNDP*** (UNDP, 2008, p. 15). Further to the above, the UNDP is subject to the United Nations Board of Auditors. This Board conducts independent audits and issues a report to the General Assembly on the audit of financial statements and other relevant financial information of the UNDP. The UNDP is also subject to oversight by the Joint Inspection Unit, which is the only **system-wide** institution mandated to conduct evaluations, inspections, and investigations. The independent Audit Advisory Committee supports the UNDP Administrator with financial management and reporting, internal and external audit issues, risk management processes, and systems of internal control and accountability. Independent internal oversight takes place through the Director of the Office of Audit and Investigations, who supports the

Executive Board in its oversight and the Administrator's accountability. The Office of Audit and Investigations also has an OAI hotline for reports of alleged breaches.

While corruption is difficult to eradicate, good governance, as the above examples illuminate, can serve as powerful, albeit, not guaranteed deterrents against corruption. When good governance is in place, by way of initiatives or 'mechanisms' that advance transparency and accountability, as examples, corruption is less likely to happen given the higher levels of accountability and transparency that can expose corruption and inefficiency. This can foster an environment that allows citizens to trust public services because good governance allows citizens to hold authorities accountable for better development results, as it encourages civic engagement in the policymaking process and increases transparency and accountability. For this reason, many international organizations have created and/or enhanced transparency and accountability mechanisms within their organizations and in their country programming activities. But there is much we need to learn about what strategies and tactics work best and why.

7.3 Accountability and Transparency Themes in Select Literature

The causes and consequences of corruption are complex. We undertook a general inquiry to find out within a targeted set of academic literature, how international organizations generally were addressing two key areas within the good governance framework; namely, transparency and accountability, was warranted. To execute this study, we turned to the framework of Arksey & O'Malley (2005) and a PRISMA flow chart to guide our review, collection, and analysis of the data.

We used three databases for our search: ProQuest, EBSCO, and Social Sciences Citation Index and eight keywords: "intergovernmental organizations" OR "international organizations" OR "international development organizations" OR "international financial institutions" OR "multilateral development banks". These keywords were combined with following keywords (**anywhere**): "accountability" OR "transparency" OR "governance". We searched articles from 1971 to present and conducted the search in February 2018. We generated 454 (ProQuest), 327 (EBSCO), and 74 (Social Science Citation Index) articles. After a rigorous selection process, we were able to identify 30 key articles and have highlighted below the core themes that emerged from them.

7.3.1 Theme 1: The History of Accountability, Transparency, and Good Governance in International Organizations

Pressure from civil society, state and non-state actors for greater accountability and transparency in international organizations prompted international organizations, largely in the 1990s, to implement mechanisms to enhance accountability and transparency within the organizations and within their country programming and/or loans. (Casaburi et al., 2000; Handl, 1998; Kapur, 2001; Thomas, 1999; Woods, 1999). These reforms aimed to achieve, "legitimacy of governments, adequate regulatory frameworks, active participation by those directly affected, capable and transparent civil service, decentralized system of policy implementation which generates accountability at all levels" (Casaburi et al., 2000, p. 495) and therein, embodies the concept good governance.

Still, a universal understanding and operationalization of what good governance meant was lacking (Kapur, 2001; Woods, 1999). As Woods (1999) points out, "much more slowly, multilateral organizations have begun to question what good governance means for the way in which they themselves are structured and in which they make and implement decisions" (p. 39). Even in the absence of conceptual clarity, good governance is in fact seen as a critical framework for more sustainable social and economic development (Kapur, 2001). According to Handl (1998), "sustainable development is achievable only in a social setting that allows for public access to information [...] and governmental accountability" (p. 651). Consequently, when there is an absence of good governance in international organizations, there is an understanding that it is linked to corruption and inefficiency (Thomas, 1999). A specific example of this is demonstrated by international financial institutions (IFIs), who have defined good governance as the "elimination of corruption through the establishment of the rule of law and through efficiency and accountability of the public sector" (Thomas, 1999, p. 552).

7.3.2 Theme 2: Accountability and Transparency Challenges Within International Organizations

The implementation of mechanisms to enhance good governance brought challenges. First, accountability was not implemented uniformly across international organizations (Bradlow, 2004; Grigorescu, 2008; Raffer, 2004; Trebilcock, 2009). As noted by Bradlow (2004), "the form of accountability can vary, depending on the circumstances" (p. 405). Examples of different forms are external and internal accountability (Raffer, 2004; Grigorescu, 2008), performance and compliance accountability, bureaucratic accountability (Grigorescu, 2008), horizontal accountability (Grigorescu, 2008), financial accountability (Raffer, 2004) and managerial accountability (Trebilcock, 2009). The diversity of where (and under what contexts) accountability was applied demands different mechanisms and policies (Raffer, 2004). It is not surprising then that this has led to diversity and variation in the types of accountability mechanisms that exist across international organizations.

Another challenge was the complex and ambiguous nature of accountability (Dunworth, 2008; Nanda, 2004). Dunworth (2008) poses the question, "What, precisely, does it [accountability] mean? How does it differ from responsibility? To whom, or what, should international organizations be accountable?" (p. 873). Similarly, Nanda (2004) recommends, "First, however, we must distinguish between accountability, responsibility, and legal liability. The definitions are not clear. Second, as the mandate and the role of international organizations (IOs), and thus the range and scope of their activities, differ among them, these variables must be considered in determining their accountability. Third, we must answer broadly the question, 'accountability to whom?'" (p. 379). For example, international organizations, donors, state and non-state actors, public and private sector and civil societies can all assume a role in the implementation of projects and are all accountable on different levels making "accountability exacerbated" (Fourie, 2015, p. 99). As a consequence of these challenges, there is no universal policy framework of accountability mechanisms in international organizations (Trebilcock, 2009). However, aiming for universality in accountability in international organizations is perhaps impossible. There is "no perfect fit for all" (Nanwani, 2014, p. 250) and because international organizations differ in structure and functions, it may be "pointless to search for general principles" (Dunworth, 2008, p. 873).

Other challenges discussed in the literature highlight problems international organizations face in implementing accountability measures. For example, the involvement of law and global governance adds complexity to the implementation of accountability in international organizations. We would like to recommend reading the articles of Suzuki & Nanwani (2005) and Parish (2010) for information about the immunity of international organizations, Orakhelashivili (2005), Nanda (2004), Parish (2010) and Fourie (2015) for the legal nature of accountability and Fourie (2015) for global governance. Further exploration of these complex subjects is beyond the scope of this discussion.

Transparency

As with accountability, international organizations have not had fluid paths in their efforts to implement transparency mechanisms. Transparency is applied in many different ways and "different types of information flows" (Grigorescu, 2003, p. 646). As a consequence, transparency policies are mostly implemented in specific areas within an international organization, such as disclosure policies (Grigorescu, 2003; Roberts, 2004). However, disclosure policies and similar efforts for achieving transparency are not working properly in international organizations (Roberts, 2004). According to Grigorescu (2003) this is due to a "lack of understanding of the concept of transparency [that] has led to problems in its operationalization" (p. 653). Even though international organizations are trying to improve their transparency policies, confidentiality rules and international relations make this difficult (Nelson, 2001;

Roberts, 2004). Again, as with accountability, there is no universal policy framework on transparency, which limits comparisons across institutions. Roberts (2004) comments on this by saying: "A general principle regarding transparency could be inferred, but universally applicable tools for achieving transparency have not been created" (p. 419). Nelson (2003) and Grigorescu (2007) both agree that international organizations cannot only implement one of these concepts, because accountability and transparency are linked. Without transparency, international organizations cannot be held accountable (Nelson, 2003) and transparency is a "necessary condition, for achieving accountability of any organization" (Grigorescu, 2007, p. 626).

7.3.3 Theme 3: Evaluation Challenges

With increasing public scrutiny of their operations, international organizations have placed more effort on creating and optimizing evaluation offices for their projects and policies (Coicaud, 2016). Coicaud (2016) offers a clear view on what the purposes of evaluations are, "to measure the extent to which the projects, initiatives and policies put in place are fulfilling their goals and, more broadly, the mandates and missions of the international organizations" (p. 421). To this effect, Coicaud (2016) comments on the growing importance of evaluation at a global level, although it is a challenge to evaluate projects and policies in international organizations (Coicaud, 2016). Especially accountability and transparency policies are difficult to evaluate, because "there is no universal standard of transparency or accountability against which all international organizations can be measured" (Deshman, 2011, p. 1102). Evaluations in international organizations are missing techniques and methodologies and are "fragmented, non-comprehensive and non-integrated" (Bauhr & Nasiritousi, 2012; Coicaud, 2016, p. 423). Additionally, international evaluation offices are an internal and missing independent review (Berkman et al., 2008; Coicaud, 2016; Fourie, 2015; Grigorescu, 2008). Although investigation and evaluation offices of international organizations are trying to become more independent (Grigorescu, 2008), Coicaud (2016) addresses the need for increasingly independent evaluation offices to avoid "bias" (p. 421).

7.3.4 Theme 4: Good Governance and the Concepts Accountability and Transparency: The Influence on Corruption

Good governance and particularly its components of accountability and transparency are seen in the literature as tools that can minimize corruption (Kapur, 2001; Parish, 2010). This is discussed by Morrisey (2007), Berkman et al. (2008) and Trebilcock (2009), and to a lesser extent by Kapur (2001), Bradlow (2004) Grigorescu (2008),

Bauhr & Nasiritousi (2012) and Parish (2010). According to our findings, corruption can be eliminated by using accountability and transparency mechanisms as anti-corruption tools in international organizations (Bauhr & Nasiritousi, 2012). These mechanisms generate higher effectiveness and efficiency in projects and activities of international organizations (Grigorescu, 2007; Trebilcock, 2009). Without transparency in an international organization, corruption can "be hidden effectively from the public" (Morrisey, 2007, p. 166). Without accountability, no international organization can inform or justify their actions to the public affected or suffer punishment in the case of corruption (Grigorescu, 2008).

In a similar vein, corruption has had a major impact on the effectiveness and efficiency of international organizations (Berkman et al., 2008). According to Berkman et al. (2008), international organizations are not effectively mitigating fraud and corruption. Increasing transparency in international organizations will lead to more clarification on wrongdoings and will increase the accountability of international organizations (Morrisey, 2007). The authors also call for enhanced collaboration (between international organizations and member states); this will greatly support a more effective and efficient organization, which can facilitate the elimination of corruption (Bradlow, 2004; Morrisey, 2007; Trebilcock, 2009).

7.4 Discussion

Rose-Ackerman and Palifka (2016) emphasize the importance of improving and bettering the good governance, accountability, and transparency (GAT) mechanisms and strategies that currently exist within and across international organizations. While there is now an abundance of mechanisms that are used to facilitate GAT, the evolution of these mechanisms has not been matched with comparable development in best practices on how these mechanisms should be operating or implemented (Deshman, 2011; Roberts, 2004; Rose-Ackerman & Palifka, 2016). Instead, Weisband & Ebrahim (2007) suggest that "the analytical domains" of these concepts, such as accountability, have been eroded (p. 2). They have become nebulous concepts that suffer from ambiguity, especially as it relates to practice. In other words, there is less or almost no clarity on the 'measuring' of these concepts and respective mechanisms. Stated differently, there has been almost no progress towards understanding which GAT mechanisms work and which ones do not, and under what circumstances.

7.4.1 Evaluation Research

The fact remains that there are many unknowns regarding the GAT mechanisms of international organizations (IOs) and the impact of their implementation. To reflect other critical discussions related to this subject, "it is not clear that efforts in the name of accountability have actually achieved their purported aims" (Weisband &

Ebrahim, 2007, p. 2). Even so, GAT mechanisms are commonly espoused as solutions to improving the effectiveness, efficiency, and progress of IOs (Weisband & Ebrahim, 2007).

Yet, in some cases, we do find examples where mechanisms of GAT are potentially impeding or running counter to the very efforts and activities that they aspire to support; this holds true for international organizations (Trebilcock, 2009; Weisband & Ebrahim, 2007; Wenar, 2006) but also other types of international institutions, such as the European Court of Human Rights (Parish, 2010). Beyond these inefficiencies, our results have also shown that IOs continue to be susceptible to misconduct (Parish, 2010). The issue at hand is that in spite of existing GAT mechanisms, international organizations are embedded in a culture of "whitewashing wrongdoings" (Parish, 2010, p. 293). The criticism, then, is that GAT mechanisms are likely not working well, though their potential to be successful is certainly up for further examination and debate.

This brings us to what we believe is a natural discussion of evaluation and learning how to successfully put these concepts of GAT into action; that is, to learn more about their practical operation and "what conditions are required to make them effective" (Rose-Ackerman & Palifka, 2016, p. 450). According to Nanda (2004), "the necessary mechanisms and procedures" (p. 379) to ensure GAT is largely unknown (p. 379). Consequently, this makes it difficult to empirically determine if (or which) GAT mechanisms are working or whether new mechanisms are needed.

The lack of this information is a persistent gap, as identified by Deshman (2011). Yet, the need for evaluation is not new. In fact, it has been discussed by several scholars (Grigorescu, 2008; Woods & Narlikar, 2001), and more recently discussed by Rose-Ackerman & Palifka (2016) and Coicaud (2016). In a similar vein, we do not suggest that evaluations of this variety have not been conducted. Instead, our review of select literature has shown that there have been instances where IOs have conducted 'internal evaluations' of their respective GAT structures and operations (Kapur, 2001; Woods & Narlikar, 2001). Consequently, these evaluations have their inherent challenges of bias (Coicaud, 2016). These evaluations are further problematized by Woods and Narlikar (2001), who discussed how such evaluation work is conducted privately and is often not published or made available to external parties. Arguably, it is through the publication of these evaluations that more actors can be informed and "apply pressure for change" (p. 576). In summary, greater, transparent, and more accessible evaluation of IOs has increasingly become important.

7.4.2 *Guarding the Guard Dogs: Challenges and Opportunities*

At this juncture, we highlight that our dialogue on GAT mechanisms is aligned with Rose-Ackerman & Palifka (2016) who support the probing of inquiry on the GAT practices across IOs. This has led to the proliferation of a complex question, "who is guarding the guard dogs?" (Kapur, 2001; Rose-Ackerman & Palifka, 2016). Our findings show that this is an especially important question to be asking, where the practice of GAT by IOs is believed to be modest (or in some cases, detrimental) and thus in need of radical transformation (Roberts, 2004). Such transformation requires an evaluation and further research, although this is not without its own unique challenges.

Primarily, the attempt for evaluation of GAT mechanisms is complicated by the wide diversity in their implementation (Trebilcock, 2009). As demonstrated in the introduction to this chapter, some mechanisms are anti corruption specific, while other IOs provide broader mechanisms of accountability and transparency. Notably, there are also many different classifications for these mechanisms, which have fallen under the purview of: auditing schemes, oversight offices, performance monitoring, and safeguards, to name a few. According to Gilbert et al. (2011), the number and scope of these "non-state regulatory initiatives has increased over the last two decades" (p. 35). This has led to a cascade of mechanisms or schemes that are arguably, likely overlapping (Gilbert et al., 2011). Yet, even with the proliferation of these mechanisms, to our knowledge, there appears to be no consensus on how to systematically classify said GAT initiatives (see discussion by Gilbert et al. (2011)). Herein lies a major challenge to evaluation work, where measuring or even comparing different mechanisms becomes almost impossible, especially given that there are no universal standards/principles to utilize or reference in such an endeavor (Deshman, 2011; Roberts, 2004).

Even though our findings point to this gap, several scholars argue that there are also many questions regarding the concepts themselves (Dunworth, 2008; Grigorescu, 2003). A corollary to this evolving thesis, then, is that we may need to achieve more clarity on concepts as a first step (Dunworth, 2008; Jordan, 2005; Weisband & Ebrahim, 2007). Certainly, our findings showed that while there are similarities between these concepts, there are some differences as well. For example, there was some discussion of the different types of accountability, with popular distinctions between vertical accountability versus horizontal accountability (Grigorescu, 2008). Similarly, Rose-Ackerman (2017) recently argued that there are still ongoing debates over these terms, such as the meaning of 'good governance' and what exactly it entails.

Given the above, there are many researchers who argue that the intentions and goals of GAT mechanisms are likely not being realized (Woods, 1999). Our findings similarly suggest that such inadequacy is likely due to the lack of incentive for IOs to implement these GAT mechanisms well (Parish, 2010). These criticisms are increasingly paralleled with questions around the legitimacy of IOs. In fact, we are seeing increasing skepticism about their legitimacy, as well as the credibility of their activities (Coicaud, 2016). Interestingly, this is not a debate exclusive to IOs but has also been raised within nongovernmental organizations and international corporations as well.

We conclude that there is a need for further inquiry to probe the impacts and effectiveness of governance, accountability, and transparency mechanisms which international organizations have adopted. International accountability standards (IAS) for businesses and corporations may be a path that IOs can follow. Through the literature on IAS, we learned that implementation alone is insufficient and that an ongoing commitment of examining IAS is needed to ensure the organizations' legitimacy (Behnam & MacLean, 2011; Gilbert et al., 2011). Similarly, the fear of losing legitimacy can be positioned as an incentive for IOs; this is an opportunity to strongly commit to GAT mechanisms and avoid any major disruption of their legitimacy. Further supported by Coicaud (2016), evaluation can certainly play a role in improving legitimacy.

The agenda for GAT mechanisms has become "intertwined with the anti-corruption agenda" (Rose-Ackerman, 2017, p. 23). But perhaps erroneously, these concepts of good governance have become "a convenient euphemism for corruption" (Rose-Ackerman, 2017, p. 23). Moving beyond the focus on anti-corruption efforts, our findings provide a reminder that evaluation can serve as tools that can (and should) be used to enhance institutional development (Nanwani, 2014), enhance operational effectiveness (Suzuki & Nanwani, 2005), promote good outcomes and better practices overall (Rose-Ackerman, 2017). Fundamentally speaking, evaluation is an opportunity to strengthen GAT structures (Berkman et al., 2008), and therein, the quality of IO activities, and not solely for the sake of mitigating corruption.

At present, we have a consensus that GAT is important. Yet, the remaining question we are tasked with answering is not *whether* to have good governance, accountability and transparency, but *how*. Consequently, we must thoughtfully begin moving toward establishing a stronger understanding of what mechanisms work best in practice and where. As we seek to meet SDG #16, building our knowledge in this area is imperative.

References

Arksey, H., & O'Malley, L. (2005). Scoping studies: Towards a methodological framework. *International Journal of Social Research Methodology, 8*(1), 19–32.

Bauhr, M., & Nasiritousi, N. (2012). How do international organizations promote quality of government? Contestation, integration, and the limits of IO power. *International Studies Review, 14*(4), 541–566.

Behnam, M., & MacLean, T. L. (2011). Where is the accountability in international accountability standards? A decoupling perspective. *Business Ethics Quarterly, 21*(1), 45–72.

Berkman, S., Boswell, N. Z., Brüner, F. H., Gough, M., McCormick, J. T., Egens Pedersen, P., … & Zimmermann, S. (2008). The fight against corruption: international organizations at a cross-roads. *Journal of Financial Crime, 15*(2), 124–154.

Bradlow, D. D. (2004). Private complaints and international organizations: A comparative study of the independent inspection mechanisms in international financial institutions. *Georgetown Journal of International Law, 36,* 403.

Casaburi, G., Riggirozzi, M. P., Tuozzo, M. F., & Tussie, D. (2000). Multilateral development banks, governments, and civil society: Chiaroscuros in a triangular relationship. *Global Governance, 6*(4), 493–517.

Coicaud, J. M. (2016). Evaluation, international organizations, and global policy: An introduction. *Global Policy, 7*(3), 420–425.

Deshman, A. C. (2011). Horizontal review between international organizations: Why, how, and who cares about corporate regulatory capture. *European Journal of International Law, 22*(4), 1089–1113.
Dunworth, T. (2008). The legal adviser in international organizations: Technician or Guardian. *Alberta Law Review, 46,* 869.
Fourie, A. N. (2015). Expounding the place of legal doctrinal methods in legal-interdisciplinary research. *Erasmus Law Review, 8*(3), 95–110.
Gilbert, D. U., Rasche, A., & Waddock, S. (2011). Accountability in a global economy: The emergence of international accountability standards. *Business Ethics Quarterly, 21*(1), 23–44.
Grigorescu, A. (2003). International organizations and government transparency: Linking the international and domestic realms. *International Studies Quarterly, 47*(4), 643–667.
Grigorescu, A. (2007). Transparency of intergovernmental organizations: The roles of member states, international bureaucracies and nongovernmental organizations. *International Studies Quarterly, 51*(3), 625–648.
Grigorescu, A. (2008). Horizontal accountability in intergovernmental organizations. *Ethics & International Affairs, 22*(3), 285–308.
Handl, G. (1998). The legal mandate of multilateral development banks as agents for change toward sustainable development. *American Journal of International Law, 92*(4), 642–665.
International Financial Institutions Anti-Corruption Task Force (IFI). (2010). *Agreement for mutual enforcement of debarment decisions.* Retrieved May 29, 2018, from http://lnadbg4.adb.org/oai001p.nsf/0/F77A326B818A19C548257853000C2B10/$FILE/cross-debarment-agreement.pdf.
Jordan, L. (2005). Mechanisms for NGO accountability. *GPPi Research Paper Series, 3*(1), 1–20.
Kapur, D. (2001). Expansive agendas and weak instruments: Governance related conditionalities of international financial institutions. *The Journal of Policy Reform, 4*(3), 207–241.
Kohler, J. C. (2011, October). *UNDP (United Nations Development Programme), fighting corruption in the health sector methods, tools and good practices.* Retrieved May 29, 2018, from http://www.undp.org/content/dam/undp/library/Democratic%20Governance/IP/Anticorruption%20Methods%20and%20Tools%20in%20Health%20Lo%20Res%20final.pdf.
Morrissey, D. (2007). The fight against corruption by international organizations. *George Washington International Law Review, 39,* 165–198.
Nanda, V. P. (2004). Accountability of international organizations: Some observations. *Denver Journal of International Law, 33,* 379–390.
Nanwani, S. (2014). Directions in reshaping accountability mechanisms in Multilateral Development Banks and other organizations. *Global Policy, 5*(2), 242–252.
Nelson, P. J. (2001). Transparency mechanisms at the multilateral development banks. *World Development, 29*(11), 1835–1847.
Nelson, P. J. (2003). Multilateral development banks, transparency and corporate clients: 'public–private partnerships' and public access to information. *Public Administration and Development, 23*(3), 249–257.
Orakhelashvili, A. (2005). Institutional aspects of the accountability of international organizations. *International Organizations Law Review, 2,* 57–102.
Organization for Economic Co-operation and Development. (OECD). (2014). *The rationale for fighting corruption.* Retrieved May 29, 2018, from https://www.oecd.org/cleangovbiz/49693613.pdf.
Parish, M. (2010). An essay on the accountability of international organizations. *International Organizations Law Review, 7*(2), 277–342.
Raffer, K. (2004). International financial institutions and financial accountability. *Ethics & International Affairs, 18*(2), 61–77.
Roberts, A. (2004). A partial revolution: The diplomatic ethos and transparency in intergovernmental organizations. *Public Administration Review, 64*(4), 410–424.
Rose-Ackerman, S., & Palifka, B. J. (2016). *Corruption and government: Causes, consequences, and reform.* Cambridge University Press.

Rose-Ackerman, S. (2017). What does "governance" mean? *Governance, 30*(1), 23–27.
Suzuki, E., & Nanwani, S. (2005). Responsibility of international organizations: The accountability mechanisms of multilateral development banks. *Michigan Journal of International Law, 27*, 177–214.
Thomas, C. (1999). Does the good governance policy of the international financial institutions privilege markets at the expense of democracy? *Connecticut Journal of International Law, 14*, 551.
Trebilcock, A. (2009). Implications of the UN convention against corruption for international organizations: oversight, due process, and immunities issues. *International Organizations Law Review, 6*(2), 513–540.
United Nations. (2015). *Sustainable development goal 16*. Retrieved May 29, 2018, from https://sustainabledevelopment.un.org/sdg16.
United Nations Development Programme (UNDP). (2008, September 12). *The UNDP accountability system: Accountability framework and oversight policy*. Retrieved May 29, 2018, from https://www.undp.org/content/dam/undp/library/corporate/Transparency/UNDP%20Accountability%20framework.pdf.
United Nations (UN). (2009). *Secretary-general, in message for international anti-corruption day, cites convention as strongest way to fight graft, build integrity*. Retrieved May 29, 2018, from https://www.un.org/press/en/2009/sgsm12660.doc.htm.
Wang, H., & Rosenau, J. N. (2001). Transparency international and corruption as an issue of global governance. *Global Governance, 7*(1), 25–49.
Weisband, E., & Ebrahim, A. (2007). Global accountabilities: Participation, pluralism and public ethics. Retrieved May 29, 2018, from https://sites.hks.harvard.edu/m-rcbg/papers/seminars/ebrahim_april_07.pdf.
Wenar, L. (2006). Accountability in international development aid. *Ethics & International Affairs, 20*(1), 1–23.
Woods, N. (1999). Good governance in international organizations. *Global Governance, 5*(1), 39–61.
Woods, N., & Narlikar, A. (2001). Governance and the Limits of Accountability: The WTO, the IMF, and the World Bank. *International Social Science Journal, 53*(170), 569–583.
World Bank. (2017). *Combating corruption*. Retrieved May 29, 2018, from http://www.worldbank.org/en/topic/governance/brief/anti-corruption.
World Health Organization (WHO). (2015). *Compliance, risk management and ethics*. Retrieved May 31, 2018, from http://www.who.int/about/finances-accountability/accountability/en/.

Sandani Hapuhennedige B.Sc., B.A., MPH, is a recent Masters of Public Health graduate from the Dalla Lana School of Public Health (University of Toronto). Her professional background includes training in research and policy analysis, predominantly tethered to the subject matter of pharmaceutical policy, governance and immunization. This includes her work with WHO Collaborating Centre for Governance, Accountability, and Transparency in the Pharmaceutical Sector, where she co-led a research project under the supervision of the Director, Dr. Jillian Clare Kohler. This work is primarily focused on promoting critical appraisal and ongoing development of stronger, good governance mechanisms and strategies across international organizations. Beyond this work, Sandani's research also includes immunization policy, specifically as it relates to legal issues with vaccine safety and adverse events following immunization. She has led analysis of relevant policies on a national scale that has garnered attention within and beyond Canada. This work has bene presented at numerous scientific conferences, including Canadian Immunization Conference 2018 and the Public Health 2018 Conference via the Canadian Public Health Association. Sandani has also authored various academic papers, editorials, and briefs on the aforementioned topics. She was the recipient of the Dr. James Rossiter CHIR Award and graduated with distinction at University of Toronto in 2018. She earned her Honours B.A. in Health Studies and Honours B.Sc. in Pharmacology from McMaster University in 2014 and 2016, respectively.

Jillian Clare Kohler Ph.D. is a Professor at the Leslie Dan Faculty of Pharmacy, the Dalla Lana School of Public Health and the Munk School of Global Affairs. She is also Director of the WHO Collaborating Centre for Governance, Transparency and Accountability in the Pharmaceutical Sector. Her professional background includes in-house pharmaceutical policy work at Unicef, the World Bank, and the Pan American Health Organization (PAHO), where she did field work in a range of countries including Brazil, Bulgaria, Haiti, and Romania. Dr. Kohler's research is focused on fair access to essential medicines with a particular focus on good governance and intellectual property rights. She pioneered the methodology on good governance in the pharmaceutical system for the World Bank, which was subsequently adopted by the WHO and has been applied in over 35 countries globally. She is the author of over 100 policy papers, journal articles and book chapters on pharmaceutical policy and is a co-editor of The Power of Pills: Social, Ethical and Legal Issues in Drug Development, Marketing and Pricing Policies (2006) and Access to Medicines as a Human Right: Implications for Pharmaceutical Industry Responsibility (2012). Dr. Kohler is a technical adviser for the Access to Medicines Index and WHO Good Governance for Medicines Programme. She was a Peter and Patricia Gruber Fellowship in Global Justice at Yale University in 2013. She earned her B.A. and an M.A. in Political Science from McGill University and a Ph.D. in Politics from New York University.

Chapter 8
Medical Ghost- and Guest-Writing as Corrupt Practices and How to Prevent Them

James Crombie

Abstract This chapter analyses the corrupting influence of medical ghostwriting and its parallel phenomenon, guest-writing on medical and pharmaceutical research, focussing on important differences between ghostwriting in the ordinary sense of the word—in which the ghostwriter is basically an editorial assistant recruited by the apparent author and is subject to the latter's authority—and medical ghostwriting in which the sponsor or the sponsor's agent assumes the prerogative of choosing not only the ghostwriter(s) but also the guest-writer who is invited to sign on as principal author and researcher (and who is subject to being replaced). In addition to the misattribution of authorship and the obvious potential for funding bias, this type of situation involves a serious problem concerning proprietorship of and access to the raw data on which the conclusions are based. Under current laws (and the TRIPs Agreement of 1994), such data are considered to be the intellectual property of the sponsor and protected as a trade secret—a situation which is incongruent with the norms governing scientific inquiry and incompatible with concern for the well-being and safety of patients. Various strategies for preventing, reducing and remediating the harms associated with ghost- and guest-writing are discussed. Codes of conduct and agreed-upon best practices should continue to be followed by the editors of medical and scientific journals, including reinforcement of the requirement for declarations of conflict of interest by authors, as well as statements concerning funding, the identity of all contributors and the nature and extent of their respective contributions. Access to the raw data of published trials should be improved and in

This chapter is a re-worked and expanded version of a paper and slide presentation which was originally part of a Symposium entitled *Combating Corruption in Health Care and Pharmaceuticals* held at the Munk School of Global Affairs of the University of Toronto on 28 November 2016. See https://pharmacy.utoronto.ca/research/whocc/past-events/ and scroll down to the title. Some of the improvements to the original presentation are attributable to helpful suggestions and comments by professional ghostwriters who were in the audience.

J. Crombie (✉)
Faculty of Health Sciences, Dalhousie University, Halifax, Canada
e-mail: ejamescrombie@gmail.com

Department of Human Sciences, Université Sainte-Anne, Pointe-de-l'Église, NS, Canada
e-mail: james.crombie@usainteanne.ca

K. T. Çalıyurt (ed.), *Integrity, Transparency and Corruption in Healthcare & Research on Health, Volume I*, Accounting, Finance, Sustainability, Governance & Fraud: Theory and Application, https://doi.org/10.1007/978-981-15-1424-1_8

particular, should cease to be impeded by laws protecting intellectual property. The sanctioning and remediation of the harm caused by misleading articles subsequently discovered to have been ghostwritten can be affected by the use of both criminal and civil law. Universities, research institutions and faculty unions should assume their responsibilities in ensuring that guest-writing is not rewarded by promotion and that fraud, when it occurs, is associated with appropriate sanctions. The regularization of medical writing as a recognized profession as well as proposals for the detection and retraction of undeclared ghostwritten material in journal archives are also discussed and it is suggested that it is not desirable that medical research and continuing medical education should be the responsibility of the same organizations which are concerned with the manufacture and marketing of medical devices and pharmaceutical products.

8.1 Introduction

Medical ghostwriting and its parallel phenomenon, *guest*-writing, involve the production—by persons directly or indirectly in the pay of a *sponsor*—of publishable articles which are subsequently signed by and attributed to what is often referred to as a *guest author*, whose scientific or academic credentials enhance both the article's credibility and its chances of being published in a reputable medical journal. This chapter attempts to do essentially three things: (1) draw attention to the ethically important differences between what may be termed 'ordinary' or 'classical' ghostwriting, on the one hand, and its medical cousin, on the other; (2) show how medical ghost- and guest-writing—as well as the sponsorship of these practices—can be seen as forms of *corruption*, which is to say that they amount to the solicitation, acceptance or provision of an illicit advantage in exchange for the faulty performance or non-performance of a social or institutional duty; (3) examine strategies which have been devised and implemented with varying degrees of success in order to limit these corrupt practices and the harms which they occasion. Typically—and here we begin to understand the ethically problematic character of the phenomenon in question—the so-called *guest author* may not only have been relieved of much of the work of wordsmithing and formatting, but will also often not have been more than tangentially involved in the design, the conduct or the supervision of the actual research, which will often have been done by other persons in the direct or indirect pay of the sponsor. The article which the guest author signs may be one of several, part of an organized campaign whose overall purpose is to promote the use of the sponsor's product. The real work, so to speak, may have been done by employees working directly for a pharmaceutical firm or, if the work has been contracted out, for what is called a clinical research organization (CRO) (Healy, 2012: 99). The direct employer of the ghostwriter or ghostwriters is often a medical communications or public relations firm. Medical ghostwriting is thus more than mere writing, and considerably more than mere wordsmithing—hence the term 'ghost management' coined by Sergio Sismondo (Sismondo, 2007; Sismondo & Doucet, 2010). Medical ghostwriting is thus to be sharply distinguished from the mere hiring of one or more editorial assistants. Unlike the ghostwriter enlisted to produce the autobiography of a semi-literate celebrity, a medical ghostwriter is typically *not* in the employ of the apparent author but is directly or indirectly in the employ of a *sponsor* who has a vested interest in the outcome of the study. The sponsor, moreover, will typically have a role in choosing the apparent author. (We will return to this question of the subordination of the apparent author to the sponsor.)

Medical ghostwriting—when twinned with guest-writing—is an essentially corrupt process involving the contamination of the ostensible purpose of a scientific study (namely the disinterested pursuit of knowledge and concern for human well-being) by an extraneous objective (namely the marketing of a product). Medical ghostwriting involves the disguising of in-house, self-interested research as research done by independent researchers with scientific credentials. We may speculate that the resulting partiality is a causal factor in the phenomenon known as "funding bias". (On the latter, see, for example, Lexchin, 2012.)

8.2 The Problem of Access to Data

The ethically and scientifically problematic nature of medical ghostwriting and guest-writing becomes especially apparent when we realize that the stated author[1] of a ghostwritten study may—or may not—have had access to the raw data on which the research is based. The stated author is, in addition, frequently not in a legal position to grant access to this data to other researchers who may be interested in re-doing the analysis. This is because the raw data is considered to be proprietary and is protected under current law as a trade secret. Having paid for the study, having contracted it out or having had its own employees do most of the work involved, the sponsor claims legal ownership of the resulting data. Ownership conveys the right to grant or to refuse access. This type of claim to proprietorship over data has been reinforced worldwide in the wake of the adoption, in 1994 of an important international treaty known as the TRIPs Agreement.[2] The TRIPs Agreement is one of the founding instruments of the World Trade Organization (WTO). It extends not only to copyrights and patents—an important consideration in the pharmaceutical context, but one which is not addressed

[1]A number of terms are used, in various contexts, to designate the person or persons whose name appears as the author of ghostwritten articles: 'guest author', 'apparent author' and 'stated author' are the most frequent. In view of the suggestion, in our title, that the activity involved may be termed 'guest-writing'—although admittedly the amount of actual *writing* may be slight—the actor may be designated by the term 'guest-*writer*'. Yet another term, proposed *infra*, is '*prête-nom*'—reflecting the fact that what the apparent author contributes is first and foremost a *name*.

[2]TRIPs Agreement: *Agreement on Trade-Related Aspects of Intellectual Property Rights* (1994). The TRIPs Agreement requires member countries of the World Trade Organization (WTO) to include protection for intellectual property rights in their respective national legislations. This protection extends not only to protection for patents on pharmaceutical products—which is a controversial aspect of intellectual property rights not addressed in the present chapter—but also to trade secrets and other types of documents considered to be private property. See WTO, *Intellectual property: protection and enforcement* at www.wto.org/english/thewto_e/whatis_e/tif_e/agrm7_e.htm (accessed 30 September 2018).

in the present chapter—but also to trade secrets and to the internal documents of business organizations. The TRIPs Agreement requires that these and other forms of 'intellectual property' shall be protected under law by the member countries of the WTO. One of the paradoxical features of the 'protection' provided by the current legal framework, however, is that the 'apparent authors' of a sponsored study are not considered to be the legal owners of one of the most important types of intellectual property associated with what purports to be 'their' research and are thus not in a position to allow other researchers to have access to the data on which a sponsored study is ostensibly based. This situation is inimical to the practice of genuine science. It is also harmful to public health. As David Healy puts it, 'without accessible data, these trials [have] the appearance of science but [are] no longer science' (Healy, 2012: 98).

The importance of access to data—especially in the case of studies in which ghostwriting turns out to have been involved—is illustrated by the case of *Study 329* and the subsequent *Study 329 Restored* in which the same data—or at least as close to the same data as it was humanly possible to obtain—were used at an interval of a dozen years by two different groups of researchers in order to come to contrary conclusions concerning the efficacy and (more importantly) the safety of off-label use in children of the antidepressant paroxetine (marketed under the brands Paxil and Seroxat).

The original Study 329 was funded by SmithKline Beecham, subsequently GlaxoSmithKline. The original article (Keller, et al., 2001) was published in July 2001 in the *Journal of the American Academy of Child and Adolescent Psychiatry (JAACAP)* under the title 'Efficacy of Paroxetine in the Treatment of Adolescent Major Depression: A Randomized, Controlled Trial'. It was signed by Martin B. Keller of the Department of Psychiatry and Human Behavior of the Brown University School of Medicine along with 21 other coauthors. The article, however, was heavily ghostwritten, although it was not presented as such in its original published form. In the bottom right-hand corner of the first page, the authors recognize the unspecified 'contributions' of a number of other individuals with the additional mention that 'Editorial assistance was provided by Sally K. Laden, M.S.' It subsequently transpired that Laden—who worked for the medical communications firm Scientific Therapeutics Information, Inc. (STI)—could perhaps more accurately be described as the originator and organizer of the publication. Her 'Project for a Journal Article' (Laden & Romankiewicz, 1998) mentions that 'STI editors will write up to six drafts of the manuscripts with the sixth being the journal submission draft' with an estimated cost for what she described as 'editorial development' of $16,750. Another document—addressed to the first-listed of the apparent authors and signed on behalf of Sally Laden by Pretre (2001) is a covering letter accompanying the delivery of final copies of the article (on paper and on diskette), ready for submission to the journal. In her signature, Pretre indicates her status as that of 'Copyediting Assistant'.

Here now, in its entirety, is the 'conclusion' of Study 329, according to the ghostwritten article published in 2001: 'The findings of this study provide evidence of the efficacy and safety of the SSRI, paroxetine, in the treatment of adolescent depression. Additional studies are called for to define the optimal length of therapy and dose of SSRIs in this population.' In Study 329 Restored, Le Noury et al. (2015)

conclude, on the contrary, that, in the same Study 329, 'neither paroxetine nor high dose imipramine showed efficacy for major depression in adolescents, and there was an increase in harms with both drugs.' Among the 'harms' identified we find suicidal ideation and attempts at suicide—which in the 2001 article were somewhat disguised as 'emotional lability'. Le Noury et al. continue: 'Access to primary data from trials has important implications for both clinical practice and research, including that published conclusions about efficacy and safety should not be read as authoritative. The reanalysis of Study 329 illustrates the necessity of making primary trial data and protocols available to increase the rigour of the evidence base' (ibid.). To this end, the apparent author of a study should always be the real owner of its data with the authority to grant access to the latter for the purpose of reanalysis. It can be strongly argued that this lack of authority over the study's data on the part of the apparent author is a more serious ethical and scientific issue that that of plagiarism and misattribution of authorship (Matheson, 2016). Misattribution of authorship is nonetheless not a negligible problem and is discussed in the following section.

8.3 Misattribution of Authorship

Ghostwriting and in particular guest-writing obviously, involve a form of plagiarism—i.e. the giving or the taking of undeserved credit for intellectual or artistic work. Clinicians, other researchers, journalists, patients and members of the general public who are the ultimate consumers of ghostwritten studies—often via paraphrases in the popular media of overenthusiastic abstracts—are entitled to be informed of the true extent of the involvement in their production by in-house research staff of pharmaceutical companies and by ghostwriters directly or indirectly remunerated by the manufacturers of the product which is being evaluated. There is also, obviously, the additional problem that the apparent author may indeed be claiming undue and exaggerated credit for the design of the study, for the research and for the word-smithing of the article; the ghostwriters may indeed be considered to be themselves the victims of a form of injustice if, as often occurs, their sometimes considerable contribution to the overall project is not sufficiently recognized.

8.4 Corruption and Illicit Advantage

The above-discussed problems—access to data, misattribution, plagiarism and fraud—are parasitic upon the fact that what we have here is a case of *corruption.* All four members of the quartet constituted by the ghostwriter, the guest author, the sponsor and the medical communications firm (if present) are complicitous in a practice in which the ostensible purpose of medical science—namely the discovery of truth and the determination of best practices in areas relevant to human health—is contaminated by the acquisition or the attempt to acquire an illicit advantage.

In the case of the apparent author of a ghostwritten study, the illicit advantage obtained is first and foremost (but not limited to) the addition of undeserved lines to a *curriculum vitae* (cv or professional resume), contributing to career advancement, as well as the possible acquisition or maintenance of the status of *key opinion leader*.[3]

The illicit advantage acquired by the sponsor of a ghostwritten study is that the latter becomes a marketing tool—not legitimately but on the basis of at least two kinds of false pretense. The first false pretense is of an epistemological nature and involves a claim to a higher degree of scientific validity than may be justified—as well as a higher probability of mistaken conclusions than might otherwise be expected. The conscious intent to deceive may be absent—or present but rationalized as 'putting the best light on things'. The second kind of false pretense involved here is attributional and involves hiding (or de-emphasizing in a misleading manner) the fact that the study being reported upon has been produced by the manufacturer of the product being studied. This second kind of false pretense can also be characterized as a failure to declare a conflict of interest.

The medical communications firm involved in the study, finally, is the paid accomplice, advisor and assistant of the sponsor. The services provided extend typically not only to the recruitment of wordsmiths but also to the identification of potential key opinion leaders (KOLs) and the recruitment of guest-writers. See the services advertised by the medical communications firm DesignWrite (2016).

As for the remuneration received by in-the-shadows wordsmiths and project-designers, the advantage received in the form of remuneration is not *necessarily* illicit, legally or even ethically, at least not to the extent that there is present a sincere belief that one is doing honest and useful-work for pay. Innocence in this regard requires a lack of intention to mislead and a lack of awareness of any intention to mislead or deceive on the part of the other members of the quartet. One can think, for example, of numerous possible scenarios where the ghostwriter has not (yet) been instructed to produce a new version of an article in order to de-emphasize unwanted side-effects of the sponsor's product—or where the ghostwriter has not yet been mysteriously instructed to avoid all direct contact with the so-called 'guest-writer'. A medical ghostwriter can be under the sincere impression of working for the advancement of medicine and the greater good of humanity. As David Healy remarks (2012), many medical ghostwriters are talented and competent young women, well-trained in science, who choose this line of work for various reasons, including the possibility of working from home. Complete absolution from complicity on the part of the wordsmith must however be tempered by respect for the civic duty of being minimally informed of the growing literature dealing with medical ghostwriting and its associated potential for causing harm—which is becoming increasingly unlikely in the case of a writer who must seek inspiration in the major medical journals. This being said, there is nothing inherently blameworthy in formatting references and bibliographies and in doing literature searches for pay.

[3]See *infra* for further remarks on the advantages which accrue to an individual considered to be a key opinion leader (KOL). See also Moynihan (2008).

The sponsors and the sub-contracted designers of organized campaigns of so-called 'ghost management' cannot, however, pretend to the innocence which attaches to many well-meaning ghostwriters—not, in any case, where a ghost-management campaign involves exaggerating the benefits or glossing over the harms associated with the use of a product. The conscious and deliberate use as a marketing tool of what is presented to the world as disinterested research is clearly a form of corruption.

8.5 The Medical Ghostwriter Is not the 'Editorial Assistant' of the Apparent Author

It is interesting at this point to reflect on some essential differences which exist between the classical case of ghostwriting—in which for example an overly busy or semi-literate celebrity seeks editorial assistance for the task of producing what purports to be an autobiography—and the structure of the relationship involved in producing a ghostwritten article in a medical journal. In the classical case, the apparent author will typically recruit the editorial assistant. In medical ghostwriting, on the other hand, it is the sponsor or the sponsor's agent who will recruit the ghostwriter—and who will also invite an 'apparent author' to 'sign on' as the author of research which others have conducted. It is thus an illusion, in the case of medical ghostwriting, to think of ghostwriters as providing mere 'editorial assistance' to the apparent author, except in those rare cases where the ghostwriter is either remunerated directly by the researcher, or directly out of the budget for a project under the direct control of the researcher. The typical medical ghostwriter is, on the contrary, as we have already observed, not employed by the apparent researcher (apparent or stated author) but rather by a medical communications firm under contract to a sponsor.

Who may fire whom? Asking this question further reveals the important structural difference between the classical case of a ghostwritten autobiography and medical ghostwriting. Whereas in the classical case, the ghostwriter may be dismissed or replaced by the apparent author, in the case of medical ghostwriting, it is the other way around: it is the apparent author who may be dismissed or replaced by the sponsor or the sponsor's agent, as is illustrated by a case reported by Barnett (2003), public affairs editor of *The Guardian*, as described to him by the eminent psychiatrist David Healy. This case is also described in some detail by Howard Brody (2007: 131). The date is 1999 and the occasion is an upcoming conference on the drug milnacipran to be held in London, including a planned company-sponsored panel discussion. The plan was that the contributed papers were to be published in a company-sponsored supplement to the *International Journal of Psychiatry in Clinical Practice*. Healy had already agreed to participate in the event 'and planned to begin work on the paper' (Brody, ibid.) on the basis of an email addressed to Healy by the scientific director for worldwide operations of a pharmaceutical firm. This email, as quoted by Brody, reads in part: 'In order to reduced [*sic*] your workload, we have had our ghostwriters produce a first draft based on your published work. I enclose it here as an attachement [*sic*] […].' The author of the email invited Healy to 'read it through and make whatever corrections you consider appropriate' (quoted by Brody, ibid.). Healy sent back some initial suggestions and produced his own draft. The sponsoring company's reaction to this was interesting. It did not refuse to publish Healy's new version of the original 'draft'—what it did, however, was to publish Healy's version separately and to have the original version of the ghostwritten text 'authored' by another specialist who read it out at the panel session and who signed the printed version. This case of one apparent author being replaced by another is also described in the report published in *The Guardian* (Bosely, 2002). The same report also describes another case in which Healy and his colleague Richard Tranter were sent an unsolicited 'first draft' of what was to be their presentation at a symposium. This 'draft' was said to have been produced 'by a medical writer from an agency which was organizing the supplement' and was accompanied by the invitation 'to edit it in any way you choose'. In response to this invitation, Healy and Tranter did in fact introduce some important qualifications—in response to which the organizers introduced their own, further modifications, to which latter Healy took exception and withdrew his name from the article (Bosely, 2002). One of the conclusions we may draw from these examples is that, clearly, the process of medical ghostwriting involves more than mere editorial assistance.

It is interesting to note, in this context, that whereas, in the classical case of a ghostwritten autobiography, a possible motive for releasing the ghostwriter from his or her function is to protect the reputation of the apparent author. In these cases involving David Healy, however, it is clear that it is the reputation of the sponsor and the sponsor's product which is the primary concern, not the reputation of the apparent author. In this respect, it is also interesting to consult the testimony of (Dr.) Mittleman (2006) who was one of the key ghostwriters called to the witness stand in the context of the Prempro Products Litigation over Hormone Replacement Therapy (HRT or, simply, HT) as documented by Fugh-Berman (2010).

Another important difference which can be observed between the situation of the classical ghostwriter and that of a professional medical ghostwriter is that, contrary to the sort of close collaboration which might be expected in the classical case between the ghostwriter and the apparent author of an autobiography, professional medical ghostwriters are often strongly discouraged by the medical communications firms which employ them, particularly the larger ones, from engaging in any kind of interaction with the 'apparent author' of the study they are working on (Logdberg, 2011).

8.6 The Guest Author Is Essentially a *Prête-Nom*

In an article published in *PLOS Medicine*, Adriane J. Fugh-Berman describes how one large medical communications firm managed an elaborate campaign in which 'dozens of ghostwritten reviews and commentaries published in medical journals and supplements were used to promote unproven benefits and downplay harms of menopausal hormone therapy (HT), and to cast […] competing therapies in a negative light' adding that 'specifically, the pharmaceutical company [name omitted] used ghostwritten articles to mitigate the perceived risks of breast cancer associated with HT, to defend the unsupported cardiovascular "benefits" of HT, and to promote off-label, unproven uses of HT such as the prevention of dementia, Parkinson's disease, vision problems, and wrinkles' (Fugh-Berman 2010).

In order to keep up appearances, as we have seen in the above-described cases involving David Healy, the so-called 'guest author' may be somewhat hypocritically encouraged to suggest modifications (which, if proposed, will not necessarily be agreed to) and to engage in some symbolic supervision of the project described in the manuscript provided. Essentially, the service which the guest author provides is not that of providing supervision and revision, but that of lending his or her name and reputation to the project. In this connection, we are reminded of the French term '*prête-nom*'. A *prête-nom* is, precisely, someone who 'lends' his or her name. In some contexts, the term '*prête-nom*' can be translated by the English word 'proxy'—and vice versa. In a well-known strategy sometimes resorted to in order to circumvent legal limits on individual political campaign contributions, for example, contractors seeking political favour have been known to provide money to their employees with the understanding that each employee should make an individual contribution in the same amount. In this way, the employees of the contractor become *prête-noms* by attaching their names to the donations, thereby disguising (but not removing) the illegal intent of giving a larger-than-legal donation to a political party. Similarly, when a medical guest author lends his or her name to a study he or she has not conducted, he or she is disguising the fact that the study is basically a promotional tool produced by or on behalf of a pharmaceutical company. Furthermore, as more and more cases of ghostwritten studies come to light—especially ones associated with products which, in spite of initial enthusiasm, are later withdrawn from the market because of undesirable side-effects—there is a cumulative effect of discrediting medical science (Gagnon & Sismondo, 2012) and the further effect of placing clinicians and patients in situations where it is difficult to know what is genuinely in the best interest of the patient.

8.7 Betrayal of Trust

In the words of Aubrey Blumsohn:

> Readers of legitimate science expect that stated authors are truly the authors, that they vouch for the work, and that they would be able to defend their findings if challenged. They expect that authors have seen and scrutinized raw data, and would be able to provide that data if asked. That it is necessary to write this indicates how much we have lost. (Blumsohn, 2006)

A key idea in the above quotation is that legitimate practitioners of science are expected to be able to 'vouch for their work' and that it is, therefore, expected not only that they have 'seen' the data but that they could 'provide that data if asked' Where these conditions are absent—and they are notoriously absent in ghostwritten studies where the data belong not to the apparent researcher but to the employer of the ghostwriter—it is a betrayal of trust to sign off as the principal author and researcher.

The act of signing on as an apparent author and *prête-nom* of a study which one has not actually conducted and to which one does not own the data is an example of corruption since it involves, in exchange for identifiable illicit advantages, a betrayal of trust. In particular, it involves a betrayal of the trust which normally attaches to a socially recognized position in society, marked and symbolized by the apparent author's university degrees, professional status and (where applicable) his or her current title within an institution devoted to advancing science and human welfare.

In the paradigmatic case of corruption, a public official accepts to depart from the strict path of duty in exchange for some kind of illicit advantage, such as a bribe. In the case of the guest author, we may ask, what are the illicit advantages which may be received in exchange for departure from what ought to be the norm? The first and most obvious advantage which accrues to the guest-writer is, as already observed, the addition of lines to the *curriculum vitae*, which are necessary for the maintenance and advancement of status within academe. It is probably quite rare that sums of money are actually paid to guest authors by either a pharmaceutical firm or by the medical communications organization which produces the article to be signed. What is frequent, however, is for the apparent authors of ghostwritten studies to accede to the other perks and privileges associated with being considered as a 'key opinion leader' or KOL by a major organization concerned with the promotion of a particular type of therapy. Key opinion leaders will be approached not only to sign off on ghostwritten studies but also to speak at conferences (expenses paid) and at events associated with continuing medical education (CME). In the latter case, in addition to the reimbursement of expenses, a fee may be paid—and this is a practice that is currently neither illegal nor widely decried ethically—at least not in mainstream opinion—as involving a conflict of interest. KOLs in receipt of such payments typically express the belief that their personal objectivity is not imperilled by accepting them (cf. Goldman 2017: 16). This is in spite of the fact that to dwell unduly, as part of one's educational presentation, upon the adverse side-effects of one's sponsor's products could be thought of as akin to biting the hand that feeds one. This situation is perhaps tolerated because the alternative to funding of CME by the private sector is funding by some other means such as direct tuition fees, subsidies from professional organizations financed by membership dues, license fees or subsidies by government financed by taxes—all of which involve apparent financial pain in at least the short term.

8.8 The Recruitment of Key Opinion Leaders (KOLs)

In this connection, it is instructive to read Dr. Brian Goldman's account (2012 and 2017) of how he found himself engaged in a career as a paid speaker and how, because of (or in spite of) his expertise in pain killers, drug diversion and addiction, he found himself 'recruited by Purdue Pharma Canada to develop and teach a curriculum on safe and responsible prescribing of opioid drugs to patients with chronic pain' (2017: 12). Goldman writes: 'I wanted to help doctors who were getting very little training in pain management and to help patients who were in pain. I never under-played the risk of addiction, always urging colleagues to carefully assess patients and to make certain other remedies had been tried first' (2012; quoted 2017:13). Goldman recognizes, however, in retrospect, that he had unwittingly become part of a campaign based on what he now considers to be 'flimsy scientific evidence' to convince physicians that 'opioid analgesics could benefit patients with chronic pain without causing addition' (2017:13). It now seems clear, in retrospect, that the danger of causing addiction through the incautious or inappropriate prescription of opioids was severely underestimated or unduly glossed over in many relevant studies (such as, among many examples, Moulin 2002)—and that the current crisis involving opioids is one of the results of this mistake. The point to be retained here, however, especially concerns not only the particular case of opioids but in general the role of physicians and academics who are recruited as KOLs.

8.9 A Special Duty

A person who, in our society, has a recognized position as a practitioner of scientific research has what we might term a *special duty* with regard to truth-seeking—in particular if the research in which the practitioner engages (or is presumed to engage) concerns medicine and health since, in that case, the special duty with regard to truth-seeking is reinforced by the general duties of non-malfeasance and beneficence associated with the Hippocratic tradition and with the concern of the medical profession for the best interests of patients (individually and collectively). This obviously applies to clinicians—but also, in the present writer's opinion, to non-clinicians when the announced subject of research is related to medicine and health. The stated authors of such studies should not claim to have played a greater role in the study they attach their name to than the role which they actually played. And they are particularly responsible for insisting that the contribution of editorial research assistants is accurately recognized and for correctly identifying their financial relationships and possible conflicts of interest. (The responsibilities of editors and managers of journals in this regard are dealt with *infra*.)

In the case of researchers with *unrecognized* positions, such as ghostwriters, editorial assistants, and research assistants, it can be said that they do not have duties attendant upon a status which they do not have—but they do, as human beings and rational agents, have—although in a weaker form than that of recognized researchers and clinicians—an identifiable duty to refrain from inflicting unnecessary harm on patients and to refrain from complicity in conspiracies to mislead. Those who make a living providing this type of service should pay particular attention to the delicacy of the situation where their remuneration and supervision is *not* insured by those persons who will subsequently (and often misleadingly) be identified as having directed and designed the study. The existence of professional associations of medical writers with codes of ethics is at least a potential source of protection in this area (AMWA, 2008, 2009).

It must be recognized that the livelihood of a large part of the population—outside of the academe and outside of the pharmaceutical industry—is dependent on salaries paid by profit-making organizations. It is a commonplace not to be disputed here that in virtue of the employee-employer relationship, both employers and employees have special duties with regard to each other. The employees of pharmaceutical companies and medical communication firms, like other types of employees, have a certain duty of loyalty towards the organization for which they work. But, as human beings, they also have a basic human duty to refrain from inflicting unnecessary harm and from complicity in conspiracies to mislead. (We mention in passing that this also holds for management personnel, owners and investors.) These duties, it can be argued, are perhaps stronger in the health-related industries than in the rest of the economy.

8.10 In Conclusion: What Is to Be Done?

The governing bodies of universities and other research institutes have, as a consequence of the expectations generated by their assumed social role, a duty to discourage scientific misconduct and to impose appropriate sanctions when their employees commit infractions of the basic norms and expectations (written and unwritten) associated with the practice of science. Where these infractions involve serious dereliction of duty (involving culpable intent, fraud or serious negligence), the sanctions for individual researchers should extend more often than they currently do so to termination of employment and/or ineligibility for further research grants. In addition, once a university or a research organization has determined in retrospect that an externally sponsored study has been guilty of serious ethical and methodological failings attributable to the influence of the sponsor, the sponsor should be disqualified from future sponsorship of studies within the university or research organization where the faulty study took place. The management and administrators of universities and other research organizations should discourage the involvement of their researchers in research concerning the health benefits or health dangers of various products when that research is financed by the manufacturers of those products. Not all 'partnerships' are necessarily good things. As is argued by the Spanish oncologist Carlos González (quoted by Lucio, 2018; cf. Rey-López & González, 2018), no one today would want to have a conference on cancer financed by the tobacco industry and the same should be the case for research in other sensitive areas. More generally, research should not be financed by for-profit organizations which are clearly in a position of conflict of interest with regard to the possible outcomes of the research.

The editors of medical and scientific journals, for their part, have their own responsibilities. The major journals have already developed codes of conduct for editors and authors favouring better disclosure of authorship and conflicts of interest (COPE, 2011; ICMJE, 2015; Kleinert & Wager, 2011a, 2011b). Authors of studies are however still not required to certify, as a condition of publication in a journal, that they have access to the raw data and that they are authorized and willing, under appropriate conditions, to allow other researchers to access the data in question. In addition to these measures setting standards for the acceptance of future articles in medical and scientific journals, attention is also increasingly directed towards the detection and reanalysis of studies in the already existing literature which—because they may have been heavily 'ghost-managed' or for other reasons—present a potential for bias and error which was not appreciated at the time of publication. The program RIAT—for 'Restoring Invisible and Abandoned Trials'—announced in Doshi et al. (2013)—of the BMJ *British Medical Journal* is an example of what can be done and is in the process of being done. Study 329 Restored (considered above) was the first project undertaken in this initiative.

Another possible change in the legal landscape—involving perhaps a lesser degree of institutional upheaval than a return to an emphasis on the public funding of research recommended in the preceding lines—is the idea that the way should be opened for those who have been harmed by ghost-managed studies to seek redress for damages under the laws of tort and civil liability—not only from the manufacturers of the products which harmed them but also from the 'guest authors' who fraudulently represent themselves as providing scientific warrant for studies which they not only did not actually design and supervise but which they could not honestly vouch for as having genuine scientific value. This proposal has been put forward by Stern and Lemmens (2011). Not only individual researchers could be targeted by such action but also the universities and research organizations which employ them, on the basis of their failure to maintain acceptable standards of conduct among their researchers. One could argue, from a prudential point of view, however, that those who have been damaged by the promotion of products by ghostwritten studies are likely to do better by themselves by going after the manufacturer than by going after the guest-writers and their universities, since the manufacturer is likely to have deeper pockets than most individual researchers—and than most universities for that matter (with a few exceptions). From the point of view of deterrence, however, the threat of civil legal action may have a greater effect on a researcher tempted by the perks associated with engaging in guest-authoring than it will on a manufacturer tempted by the increase in sales to be obtained by a successful campaign of ghost management, since manufacturers can write off the cost of defending themselves against such claims as a cost of doing business—as they seem to do with the fines they often pay for other types of infractions concerning the illegal promotion of their products for off-label uses. In view of the deterrence-value of civil action targeting guest authors, it might be justified to provide some kind of legal aid to those victims of harm who might consider taking action against the guest writers of a misleading ghostwritten study.

In the meantime, health professionals should be very careful in evaluating the information and advice they receive through various channels, including publications in scientific and medical journals–in the light of the revelations of fraud and conflict of interest which have already come to light—and they should always seek out independent sources of information and data, especially on the newer therapies still under patent and for which a body of clinical experience has not yet had time to accumulate. Advice provided by organizations such as Prescrire.org (with a journal and other publications in both French and English)—which latter relies on subscriptions and is independent of both commercial and government funding—is available and oriented towards health care professionals. The Cochrane Collaboration prides itself on its independence—but mainly engages in meta-analyses of the best-quality studies which may, in the end, however, turn out to have been ghostwritten. There have been various attempts to constitute databases of adverse events and undesired side-effects associated with various products, including information available (some of it for free and some by subscription only) via the Best Pills Worst Pills initiative of Public Citizen in the United States. More radically sceptical with a decidedly anti-big-pharma animus, are the Spanish-language services associated with the No Gracias movement in Spain and the Latin American Social Medicine Association (medicamentos.alames.org).

Ordinary citizens, legislators and governments, as well as the regulatory agencies and other organizations they create, including international and intergovernmental institutions like the World Health Organization and the World Trade Organization have a duty to institute and to enforce treaties, laws and regulations which provide appropriate incentives for research in the area of health care and medicine. Appropriate incentives are those which encourage useful and necessary research in priority areas—but do not incentivize fraud and abuse. What appropriate incentives would should look like, exactly, is a matter for debate—but the prevalence of the more corrupt forms of ghost- and guest-writing in medical and scientific journals is a strong indication that the current structure of incentives is far from optimal. One of the main weaknesses of the system where much research turns out to have been privately produced under contract and written up by ghostwriters in order to be signed by guest authors is that this system favours lack of access by other researchers to the raw data which is the ultimate justification of whatever conclusions may have been announced in the resulting article (and its abstract). One needed reform in national and international law would be to insure that the raw data used in published studies concerning medicine and health should be considered to be public (never private) property—and should be permanently available in order to be verified and reanalysed by other researchers. Access to the raw data used in studies which the authors or the sponsors decided *not* to publish (for various reasons) would also be important, but it is not clear to the author of the present lines just how this result can be obtained. This is a question more difficult to regulate than merely insuring subsequent access to data referred to by published articles in prestigious journals through written statements to the effect that the apparent author does, in fact, have the authority to grant access to the data and, in addition, provides a commitment to do so when requested.

Ordinary citizens, legislators, and governments, should also reconsider the current fashion in favor of the increasing involvement (under the buzzword 'partnerships') of what is termed 'the private sector' in medical research. Research by publicly funded, independent agencies is relatively immune from the incursion of the corrupt forms of ghostwriting, guest-authoring and ghost-management, compared to privately funded research from within the industry. Resorting to private funding results in a short-term, apparent economy in public expenditure (with a corresponding reduction in taxes in view). This overall saving for the ultimate customer is at best merely apparent, however, since the private money spent on such research must be recovered in the price of products and services marketed subsequently to the public, with or without the advantage of a monopoly price-protection provided by patents,[4] so that the ultimate saving approaches zero—or becomes actually negative in view of the fact that there will be inevitable losses due to corruption and faulty science. There is also the consideration that, when the overall research agenda is not governed by considerations of the public good but by more limited objectives, the determination of priorities for the research actually undertaken may be less than optimal from the point of view of the average citizen. We have already referred to the observation by Carlos González that most of us would easily recognize the inappropriateness of an academic conference on cancer funded by the tobacco industry.[5] The suggestion is that we should therefore also recognize the inappropriateness of industry funding for research in other 'sensitive' areas such as health care and nutrition. Where public health is at stake, the funding of research should preferably be through publicly funded, arms-length, independent agencies—and the data generated should be considered to be the collective property of the community.

References

AMWA American Medical Writers Association. (2008, June). AMWA code of ethics (Third rev.). Retrieved January 13, 2017, from http://www.amwa.org/amwa_ethics.

AMWA American Medical Writers Association. (2009). AMWA ethics FAQs. Retrieved January 13, 2017, from http://www.amwa.org/amwa_ethics_faqs.

[4]As already mentioned in a previous footnote, we have in this article avoided the controversial question of pharmaceutical patents. The existence of a patent on an invention conveys to its holder the sole right for a specified number of years to manufacture and market the invention in question, increasing the price which may be obtained for it. It is argued in the main text that the ultimate cost to the final users of pharmaceutical products developed through privately funded research will be higher than when the development takes place through publicly funded research—and that this would be true even in the absence of patent-protection. It is suggested here that the difference would of course be greater in the presence of pharmaceutical patents.

[5]The context of González's remark is the study (Rey-López & González, 2018) he co-authored on the influence of funding by Coca-Cola of studies in Spain relating to the health effects of the consumption of soda drinks. Cf. Lucio (2018).

Barnett, A. (2003). Revealed: How drug firms 'hoodwink' medical journals: Pharmaceutical giants hire ghostwriters to produce articles—then put doctors' names on them. *The Guardian*, 7 December. Retrieved October 1, 2018, from https://www.theguardian.com/society/2003/dec/07/health.businessofresearch.

Blumsohn, A. (2006). Authorship, ghost-science, access to data and control of the pharmaceutical scientific literature: Who stands behind the word? *Professional Ethics Report, XIX*(3, Summer). Washington: American Association for the Advancement of Science. Retrieved November 14, 2016, from www.aaas.org/sites/default/files/migrate/uploads/per46.pdf.

Bosely, S. (2002). It said the drug was the best thing since sliced bread. I don't think it is. *The Guardian*, 7 February. Retrieved October 3, 2018, from https://www.theguardian.com/uk/2002/feb/07/research.health.

Brody, Howard. (2007). *Hooked: Ethics, the medical profession, and the pharmaceutical industry*. Lanham, MD: Rowman & Littlefield.

CEM/CEO Collect if Europe et Médicament & Corporate Europe Observatory. (2015). *Directive européenne sur les secrets d'affaires: Menace sur l'accès aux données de santé publique* ["European Directive on Trade Secrets: A Threat to Access to Public Health Data"]. Analysis and position paper published on the website of the NGO Formindep (founded in 2004), 6 pages. Retrieved November 20, 2016, from www.formindep.org/IMG/pdf/SecretsAffaires_AnalyseFR_FINAL.pdf.

COPE Committee on Publication Ethics. (2011). *Code of conduct and best practice guidelines for journal editors*. Retrieved December 2, 2016, from http://publicationethics.org/files/Code_of_conduct_for_journal_editors_Mar11.pdf.

Crombie, J. (2016). Medical ghost- and guest-writing as corrupt practices and how to prevent them. Slides accompanying a presentation as part of the symposium entitled *Combating Corruption in Health Care and Pharmaceuticals* which took place at the Munk School of Global Affairs, Toronto, Canada, 28 November. Retrieved October 9, 2018, from http://www.pharmacy.utoronto.ca/sites/default/files/upload//who_cc/symposium/crombie_ghosts-guests-sponsors-munk-20161205a.pdf.

DesignWrite. (2016). Case studies. Tab of Corporate Website. Princeton, NJ. Retrieved November 13, 2016, from www.dwrite.com/CaseStudies.aspx.Html.

Doshi, P., Dickersin, K., Healy, D., Vedula, S. S., & Jefferson, T. (2013). Restoring invisible and abandoned trials: A call for people to publish the findings. *British Medical Journal, 346,* f2865. Retrieved October 19, 2018 from https://doi.org/10.1136/bmj.f2865.

Fugh-Berman, A. J. (2010, September). The haunting of medical journals: How ghostwriting sold "HRT". *PLoS Medicine, 7*(9), e1000335 (11 pages, including tables and references). Retrieved November 12, 2016, from https://doi.org/10.1371/journal.pmed.1000335.

Gagnon, M.-A., & Sismondo, S. (2012, October 2). Des auteurs-fantômes discréditent la recherche médicale ["Ghostwriters discredit medical research"] Opinion piece. *Le Soleil*, daily newspaper, electronic edition. Retrieved October 15, 2013, from www.lapresse.ca/le-soleil/opinions/points-de-vue/201210/01/01-4579239-des-auteurs-fantomes-discreditent-la-recherche-medicale.php.

Grant, J. (2018). *Corrupted science: Fraud, ideology and politics in science* (Revised & expanded ed.). Tucson, Arizona: See Sharp Press.

Healy, D. (2012). *Pharmageddon*. University of California Press.

ICMJE International Committee of Medical Journal Editors. (2015). *Recommendations for the Conduct, Reporting, Editing and Publication of Scholarly Work in Medical Journals*. Also known as the *ICMJE Recommendations*. Previously known as the *Uniform Requirements for Manuscripts Submitted to Biomedical Journals (URM)*. The URM was first published in 1978, 17 pages. Retrieved November 15, 2016, from www.icmje.org/icmje-recommendations.pdf.

Keller, M. B., Ryan, N. D., Strober, M., Klein, R. G., Kutcher, S. P., Birmaher, B., et al. (2001). Efficacy of paroxetine in the treatment of adolescent major depression: A randomized, controlled trial. *Journal of the American Academy of Child and Adolescent Psychiatry (JAACAP), 40*(7), 762–772.

Kleinert, S., & Wager, E. (2011a). *Responsible research publication: International standards for editors*. A position statement developed at the 2nd World Conference on Research Integrity, Singapore, July 22–24, 2010. In T. Mayer & N. Steneck (Eds.), *Promoting research integrity in a global environment* (pp. 317–328). Singapore: Imperial College Press/World Scientific Publishing. Retrieved November 15, 2016, from http://publicationethics.org/files/International%20standard_editors_for%20website_11_Nov_2011%20%281%29.pdf.

Kleinert, S., & Wager, E. (2011b). *Responsible research publication: international standards for authors*. A position statement developed at the 2nd World Conference on Research Integrity, Singapore, July 22–24, 2010. In T. Mayer & N. Steneck (Eds.), *Promoting research integrity in a global environment* (pp. 309–316). Imperial College Press/World Scientific Publishing, Singapore. Retrieved November 15, 2016, from http://publicationethics.org/files/International%20standards_authors_for%20website_11_Nov_2011_0.pdf.

Laden, S. K., & Romankiewicz, J. A. (1998). *Adolescent Depression Study 329: Proposal for a Journal Article*. Proposed to Ivan Gergel, MD, SmithKline Beecham Pharmaceuticals, 12 pages. Produced By GSK In In Re Paxil, C.P.Ct.PA (On-Drug). *Drug Industry Documents Archive (DIDA)*, University of California San Francisco Library. Retrieved October 9, 2018, from https://www.industrydocumentslibrary.ucsf.edu/drug/docs/#id=npfw0217.

Le Noury, J., Nardo, J. M., Healy, D., Jureidini, J., Raven, M., Tufanaru, C., & Abi-Jaoude, E. (2015). Restoring study 329: Efficacy and harms of paroxetine and imipramine in treatment of major depression in adolescence. *British Medical Journal, 351,* h4320. Retrieved October 9, 2018, from https://www.bmj.com/content/351/bmj.h4320.long, https://doi.org/10.1136/bmj.h4320.

Le Noury, J., Nardo, J. M., Healy, D., Jureidini, J., Raven, M., Tufanaru, C., & Abi-Jaoude, E. (2016). Study 329 continuation phase: Safety and efficacy of paroxetine and imipramine in extended treatment of adolescent major depression. *International Journal of Risk and Safety in Medicine, 28*(3), 143–161. Retrieved October 9, 2018, from https://www.ncbi.nlm.nih.gov/pmc/articles/PMC5044781/. Online: https://doi.org/10.3233/jrs-160728.

Lexchin, J. (2012). Sponsorship bias in clinical research. *JOS—International Journal of Risk and Safety in Medicine, 24*(4), 233–242. Retrieved October 3, 2018, from https://content.iospress.com/articles/international-journal-of-risk-and-safety-in-medicine/jrs574, https://doi.org/10.3233/jrs-2012-0574.

Lexchin, Joel. (2017). *Doctors in denial: Why big pharma and the Canadian medical profession are too close for comfort*. Toronto: James Lorimer & Company Ltd.

Lexchin, J., Kohler, J. C., Gagnon, M.-A., Crombie, J., Thacker, P., & Shnier, A. (2018). Combating corruption in the pharmaceutical arena. *Indian Journal of Medical Ethics, 3*(3 (NS)), 234–239. Retrieved October 3, 2018, from https://doi.org/10.20529/IJME.2018.022.

Logdberg, L. (2011, August). Being the ghost in the machine: A medical ghostwriter's personal view. *PLoS Medicine, 8*(8). Retrieved March 29, 2012, from www.plosmedicine.org/article/info%3Adoi%2F10.1371%2Fjournal.pmed.1001071.

Lucio, C. G. (2018). La mano invisible de la industria en la ciencia ["The invisible hand of industry in science"]. *El Mundo*, 8 October. Retrieved October 14, 2014, from www.elmundo.es/papel/2018/10/08/5bb75af6ca4741b33f8b45ef.html.

Matheson, A. (2016, August 30). Ghostwriting: The importance of definition and its place in contemporary drug marketing. *British Medical Journal*. Retrieved November 12, 2016, from www.bmj.com/content/354/bmj.i4578.

Mittleman, K. D. (2006). Deposition (sworn testimony) on May 03, as transcribed from a video recording by Kimberly A. Cahill, approved reporter of the United States District Court. In Re: Prempro Products Liability Litigation, Mdl Docket No. 4:03CVL507 WRW. Available via the Drug Industry Documents Archive (DIDA) created by the University of California San Francisco Library. Retrieved November 11, 2016, from www.industrydocumentslibrary.ucsf.edu/drug/docs/qxww0217.

Moulin, D. E., Clark, A. J., Speechley, M., & Morley-Forster, P. K. (2002). Chronic pain in Canada—Prevalence, treatment, impact and the role of opioid analgesia. *Pain Research and Management,*

7(4), 179–184. Retrieved October 2, 2018, from https://www.hindawi.com/journals/prm/2002/323085/abs/.

Moynihan, R. (2008, June 19). Key opinion leaders: Independent experts or drug representatives in disguise? *British Medical Journal, 336,* 1402. Retrieved November 12, 2016, from www.bmj.com/content/336/7658/1402.

Pretre, C. (2001). Letter dated 6 February addressed on behalf of Sally Laden to Martin B. Keller, M.D. Brown University School of Medicine on the letterhead of Scientific Therapeutics Information, Inc. Produced By GSK In In Re Paxil, C.P.Ct.PA (On-Drug). *Drug Industry Documents Archive (DIDA)*, University of California San Francisco Library. Retrieved October 9, 2018, from https://www.industrydocumentslibrary.ucsf.edu/drug/docs/#id=zxfw0217.

Rey-López, J. P., & González, C. A. (2018, August 29). Research partnerships between Coca-Cola and health organizations in Spain. *European Journal of Public Health*, cky175. Retrieved October 14, 2018, from https://doi.org/10.1093/eurpub/cky175.

Sample, I. (2011, August 2). Scientists credited on ghostwritten articles 'should be charged with fraud': Legal experts call for severe sanctions against scientists who 'guest author' papers written by drug companies. *The Observer*. Retrieved March 29, 2012, from www.guardian.co.uk/science/2011/aug/02/scientists-ghostwritten-articles-fraud.

Sismondo, S. (2007). Ghost management: How much of the medical literature is shaped behind the scenes by the pharmaceutical industry? *PLoS Medicine, 4*(9), e286. Retrieved December 2, 2016, from http://journals.plos.org/plosmedicine/article?id=10.1371/journal.pmed.0040286.

Sismondo, S., & Doucet, M. (2010). Publication ethics and the ghost management of medical publication. *Bioethics, 24*(6), 273–283. Retrieved November 15, 2016, from http://onlinelibrary.wiley.com/doi/10.1111/j.1467-8519.2008.01702.x/pdf.

Stern, S., & Lemmens, T. (2011, August 2). Legal remedies for medical ghostwriting: Imposing fraud liability on guest authors of ghostwritten articles. *PLoS Medicine*. Retrieved November 12, 2016, from http://dx.doi.org/10.1371/journal.pmed.1001070.

James Crombie is an associate professor of philosophy in the Department of Human Sciences of Université Sainte-Anne, a small francophone university situated in an Acadian area of Nova Scotia, Canada. He is also Adjunct Professor in the Faculty of Health Sciences at Dalhousie University (Halifax). His research focuses on the history and philosophy of science, technology and medicine in the ethical context of globalization and theories of development (with a particular interest in Latin America). This interdisciplinarity is largely determined by the fact that, as the only philosophy professor at his university, James has been called upon to teach courses in a wide variety of philosophical sub-disciplines. His master's thesis was on C. S. Peirce and his doctoral thesis dealt with the perceptual theories of the Scottish philosopher Thomas Reid. In recent years, James has been working on the effects of the global IP system (patents, trade secrets, compulsory licensing, etc.) on the health of populations and equitable access to essential medicines. He is currently researching the life and work of the Costa Rican scientist, Clodomiro Picado Twight (1887–1944).

Part III
Fraud Schemes & Cases & Fraud Prevention in Healthcare & Pharmaceutical Sector

Chapter 9
Fraudulent Misrepresentation and Fraudulent Concealment in Products Liability in Tort Law in Canada: The Special Relationship Between Drug Companies and Consumers in the Context of the Fraudulent Misrepresentation and Fraudulent Concealment of Data

Adrienne Shnier

Abstract The legal concept of the "special relationship", arising in the areas of negligence and intentional torts, has been afforded little attention in products liability cases in Canadian law. The special relationship has been applied in cases wherein plaintiffs have suffered damages in the course of their relationships of reliance with defendants. Although the special relationship does not amount to a fiduciary relationship, the special relationship triggers a positive duty owed to the plaintiff by the defendant, where the defendant makes false or misleading representations on which plaintiffs rely. Issues of limitation period expiry and discoverability are also relevant in cases, where the intentional tort of fraudulent misrepresentation of a material fact relating to known risks prevent consumers from identifying their potential for causes of action prior to the expiry of the typical 2-year limitation period. For instance, in cases where data is withheld, misrepresented or mischaracterized, and a consumer can make a products liability claim in negligence or intentional torts, the special relationship is a factor in the analysis of responsibility, duty and relationship between the consumer and the defendant. In these cases, a plaintiff may advance the equitable doctrine of fraudulent concealment. To successfully establish fraudulent concealment in equity, one requirement is that the plaintiff must establish a special relationship between the plaintiff(s) and the defendant. Both defining and establishing the special relationship in law will likely be challenging owing to the absence of a clear definition of what constitutes a special relationship in Canadian case law and legislation, in addition to limited literature, interpreting the criteria that must be satisfied to establish the special relationship. By providing an analysis of the legal

A. Shnier (✉)
School of Health Policy & Management, 4700 Keele St, Toronto, ON M3J 1P3, Canada
e-mail: adrienne.shnier@gmail.com

K. T. Çalıyurt (ed.), *Integrity, Transparency and Corruption in Healthcare & Research on Health, Volume I*, Accounting, Finance, Sustainability, Governance & Fraud: Theory and Application, https://doi.org/10.1007/978-981-15-1424-1_9

test to establish the special relationship, related case law, primary documents and secondary research, as well as a case study to which this analysis may apply, this paper provides a thought experiment of a framework and method by which a special relationship can be established in products liability cases.

Keywords Fraudulent concealment · Fraudulent misrepresentation · Products liability law · Intentional tort law · Special relationship · Canada

9.1 Introduction

The concept of the "special relationship" in Canadian tort law has been afforded little analytical attention in the case law and scholarship, despite its potential for use in products liability cases. The special relationship notion has been introduced in two central capacities in Canadian tort law in the areas of negligence and intentional torts, as well as in other areas of law. The special relationship concept has been used as a factor to consider in cases in which plaintiffs have suffered physical, psychological, emotional or economic harms in the course of their relationships with defendants and where the plaintiffs relied on the defendants' representations. Although it is clear that the special relationship does not amount to a fiduciary relationship where the duty owed to the plaintiff by the defendant is fiduciary in nature, the special relationship has been deemed to trigger some sort of positive duty to the plaintiff. This chapter attempts to determine the characteristics of the special relationship in hopes that interrogations and analyses of the special relationship will become more prominent in the scholarship and case law.

There are two particular instances in which the special relationship may be used by plaintiffs in tort law. First, fraudulent misrepresentation claims may be advanced by plaintiffs in pharmaceutical harms cases wherein manufacturers have published misleading statements or inadequate warnings regarding the safety, efficacy and side effects profiles of their drug products. These claims may also be brought in negligence for a failure to warn; however, if the act of misrepresentation can be proven to be intentional, the act rises from negligence to an intentional tort. In fraudulent misrepresentation cases, a plaintiff may not have realized that a cause of action existed based on a misrepresentation of fact that the plaintiff, once realized, believes to be intentional. It is understandable, then, that a plaintiff would not be able to bring a lawsuit within the 2-year limitation period in Ontario, for example, if the misrepresentation continued to be concealed during this time. In a case such as this, a defendant may seek to have the plaintiff's proceedings dismissed on the grounds that the limitation period had expired. The plaintiff can advance a counterargument relying on the equitable doctrine of fraudulent concealment in an effort to obtain an extension of the limitation period, for which the plaintiff must establish a "special relationship" between the plaintiff and defendant. Defining the special relationship is challenging, considering that such a relationship and its criteria have not been clearly defined in the literature, legislation, or case law.

It is plausible that, in product liability claims, the manufacturers may seek to avoid liability by arguing that there existed no special relationship between the manufacturer and consumer due to the relational distance between the two. This distance may be due to the extent to which the pharmaceutical industry's supply chain is truly global in its outsourcing of market management, research, development, manufacturing, distributing and selling of prescription drugs. An analysis of specific steps in the pharmaceutical production and management supply chain, in combination with agency theory principles in the context of pharmaceutical industry practices, may provide support for establishing the existence of the requisite special relationship between pharmaceutical companies and their end-stage consumers. Establishing the existence of this relationship could result in a judge deciding to delay, in equity, the limitation period where expiration of limitation periods has become germane to the pleadings.

Second, the plaintiff may use the special relationship in negligence claims. The special relationship is discussed in the negligence case law, where, in order to plead negligent misrepresentation, both requirements for a *prima facie* duty of care must be met. The plaintiff is first required to establish a relationship of proximity between the plaintiff and defendant, based on the facts. The plaintiff must, then, establish that the defendant failed to take reasonable care, which may foreseeably cause loss or harm to the plaintiff (*R v Imperial Tobacco Canada Ltd.* [*Imperial Tobacco*], 1972). A special relationship will be found in negligence where the defendant could have reasonably foreseen that the plaintiff would rely on the defendant's representation, and where the plaintiff's reliance on the representation would be reasonable on the facts (*Imperial Tobacco*, 1972). The case law clearly provides that there is an opportunity for plaintiffs to attempt to establish a special relationship in both areas of negligence and intentional torts.

This chapter presents an interdisciplinary thought experiment on establishing the existence of a special relationship in law between pharmaceutical companies and the consumers of their products when one or more material facts regarding known side effects have been misrepresented. This chapter first considers Supreme Court of Canada decisions on the special relationship, followed by a critical analysis of the relevant components of the pharmaceutical business structure alongside the case law to assess the evidence for establishing the existence of a special relationship between manufacturers and consumers on the facts, with reference to a case study of paroxetine (Paxil).

9.2 Case Study: GlaxoSmithKline, Paxil and the Suppression of Side Effects Data

Recent research, investigations and lawsuits have revealed cases in which pharmaceutical companies, as manufacturers of their own pharmaceutical products, have failed to release to regulators, physicians and patient-consumers important and potentially life-saving adverse events data about which the company knew prior to its market approval. In July 2012, GlaxoSmithKline pleaded guilty to criminal charges and was fined US$3 billion to resolve its fraud charges for its illegal promotion of certain prescription medications and its failure to report certain safety data (The United States Department of Justice, 2012). A key piece of evidence in these pleadings was Study 329, a 2001 clinical trial on paroxetine (brand name: Paxil; manufacturer: SmithKline Beecham, now GlaxoSmithKline) and imipramine (brand name: Tofranil; manufacturer: Mallinckrodt Pharmaceuticals), which was published in the Journal of the American Academy of Child & Adolescent Psychiatry (Keller et al., 2001). This 2001 clinical trial publication, which had data misrepresentation problems and was largely ghostwritten (Le Noury et al., 2015a, 2015b), concluded that paroxetine and imipramine were safe and effective for adolescents diagnosed with major depression. In 2013, the "restoring invisible and abandoned trials" (RIAT) initiative, launched by the BMJ and PLOS academic peer-review journals, called for access to the clinical trial data in order to correct reported studies, including Study 329 on paroxetine and imipramine. For the purposes of this chapter, only GlaxoSmithKline's conduct surrounding paroxetine (Paxil) will be considered because of the available evidence of fraud on the part of GlaxoSmithKline.

The RIAT restoration of Study 329, published in 2015 (Le Noury et al., 2015b), found that the study's original conclusions were at odds with the data. Using the original study protocol in accordance with the RIAT restoration declaration (BMJ, 2013), the RIAT team's reanalysis of the clinical trial data found that paroxetine and imipramine failed to show efficacy for the treatment of major depression in adolescents and, further, showed no clinically or statistically significant benefit over placebo in adolescents with depressive symptomatology, as defined in the study. Furthermore, the RIAT restoration found clinically significant increases in adverse events from paroxetine, including suicidal ideation and suicidal behaviours; however, the original clinical trial data showed notable data irregularities as compared with the RIAT restoration using the same protocol. The RIAT team followed the original protocol and found important non-transfer of information that extended from the participants' case report forms (CRFs), to the clinical study reports (CSRs), to the data tables that often misrepresented and re-coded serious adverse events using less serious-sounding terminology.

Appendix 3 of the RIAT restoration (Le Noury et al., 2015a) reveals the decision-making system that led to the misrepresentation of serious adverse events, which formed the statistical underpinning for GSK's assertions about the safety and efficacy of paroxetine. In Table C of Appendix 3, we see that a participant's intentional swallowing of 80 Tylenol tablets was blandly coded as "emotional lability" using ADECS coding, rather than the more accurate depiction of the serious adverse event of "suicidal ideation" or "self-harm/attempted suicide" using the MedDRA coding system (Le Noury et al., 2015a). There were several intentional overdoses, superficial cuts to the skin, self-mutilation, suicidal ideations and intentional attempted suicides that were also coded as "emotional lability", rather than the more contextually accurate coding terminology of "suicidal ideation" or "self-harm/attempted suicide" (Le Noury et al., 2015a). Another example of the misrepresentation of serious adverse events was the coding of "need[ing] 6 stitches to hand after breaking pictures (due to anger) [that] resulted in hospitalization to prevent aggression against self" as "hostility" (Le Noury et al., 2015a), rather than a more accurate term such as "self-inflicted laceration" (IFPMA and ICH, 2013).

The misrepresentation of serious adverse events data at the clinical trial level of the pharmaceutical research and development (R&D) structure offers unique insights into the points in a drug's development timeline at which key decisions shape not only a drug's safety profile and narrative, but also may indicate whether a drug company has a special relationship with consumers. The circumstance in which this chapter is interested is when key data is arguably intentionally factually misrepresented prior to its submission to drug products regulators such as Health Canada and the Food and Drug Administration (FDA) and, certainly, before it reaches physicians, academic researchers and the public. Therefore, the question of whether a drug company could be burdened with a special duty to the consumer of its medications that reaches the status of a special relationship warrants consideration in light of particular evidence. This evidence includes proof of the misrepresentation of facts on which consumers rely, the unique business structure of large drug companies, the distinguishable characteristics of prescription medications as commercial products, the significance of the ingestion or absorption of pharmaceutical products, and the relationships between drug companies, regulators, physicians, the medical literature and the ultimate consumers.

9.3 Products Liability in Tort Law

Products liability law in Canada has its foundations in tort law and, depending on the facts, is generally based on the principles of negligence or intentional torts. Harms to consumers resulting from products can result from either negligence or intentional torts. Products liability law is intended to be a method by which consumers can protect and defend themselves against harms resulting from manufacturers' potential engagement in negligent or intentional behaviours that cause risk of harm to consumers. Conversely, products liability law also functions as a deterrent to companies' participation in negligent or intentional torts that cause risk of harm to consumers. There are typically three types of product liability claims based on the tort of negligence: design defects, manufacturing defects and failure to warn by way of misleading or inadequate warnings (Harrison, 2013). The fourth type of products liability claim is based in intentional torts and includes such conduct as fraudulent misrepresentation, or a false statement of fact by the defendant, on which the plaintiff detrimentally relied (Harrison, 2013).

A manufacturer's duty to warn is continuous and does not conclude with its sale of the product to the consumer (Osborne, 2015). The manufacturer is tasked with the ongoing duty to warn the consumer of its products' inherent risks and dangers that are discovered not only prior to its sale on the market but also while its products remain on the market for sale to consumers. This principle, which encompasses the *Lambert* principle, introduced below, applies equally to medical devices and prescription drugs, which are the focus of this chapter. Pharmaceutical manufacturers and suppliers are burdened with the heavy obligation of providing clear, complete and current information regarding the dangers and risks associated with the intended and ordinary use of their pharmaceutical products (Osborne, 2015). To be successful in a products liability claim against a manufacturer for failure to warn in negligence, a plaintiff must demonstrate that there was a failure to warn and the negligence of the failure to warn was the cause-in-fact of the harm to the plaintiff. This issue of causation, at the core of every product liability claim, must be established for the court, but is outside of the scope of this chapter.

In tort law, manufacturers of products have an initial duty in negligence to take reasonable care to manufacture products, which are to be mass produced and consumed, that are free of dangerous defects (Osborne, 2015; *Donoghue v Stevenson* [*Donoghue*], 1932). This duty of care is complemented by the duty to warn principle, which dictates that manufacturers have a duty to warn consumers of the inherent dangers of their products, as well as a duty to design products with reasonable care to avoid product defects. Design defects are typically unintentional, affecting a small minority of the manufactured products. The liability on manufacturers for product defects is typically less burdensome than the liability arising from a failure to warn because product defects usually arise from mistake or malfunction.

The liability arising from a manufacturer's failure to warn consumers of the risk for harms from their products is more burdensome due to the nature of the tort and its fault requirement (Osborne, 2015). In this case, the product need not be defective, but applies where a manufacturer knew of dangers associated with a product's use and failed to warn the public of the risk for danger. The leading Canadian decision on manufacturer's duty to warn is *Lambert v Lastoplex Chemicals Co.* [*Lastoplex*] (1971), in which the plaintiff used a volatile lacquer sealer to seal his basement floor when, upon its use, caused injury by exploding when a pilot light was exposed to its fumes. The manufacturer, Lastoplex Chemicals Co., included a warning on the can for its users to avoid using the lacquer sealer near an open flame, but failed to warn its users of the requirement to extinguish all pilot lights in the vicinity of its use. The Supreme Court of Canada held that a duty to warn on the part of the manufacturer arises when a product is placed on the market for use by the general public, the intended use of the product is dangerous, and the public is not afforded the same awareness or knowledge of the danger that the manufacturer possesses.

The standard of care that is applied in products liability cases is to take reasonable steps to provide warnings to allow consumers to safely use the product in question (Stikeman Elliott LLP, n.d.). This standard of care is fact-specific, varying based on the nature and degree of danger that could arise from the use of the product, distinctiveness, intensity, clarity and extent of any written warning, as well as the standard practice of manufacturers that produce similar products (Osborne, 2015). The standard of care is also determined based on the location of the warning, for example, whether the warning is written on the product itself, package inserts that accompany the product, or in accompanying literature, as well as the nature of marketing of the product, commercial practice and the conduct of reasonable consumers. Meeting the statutory requirements for the manufacturer's duty to warn may be insufficient to protect the plaintiff from the risks associated with the product (Osborne, 2015). For instance, in *Lastoplex*, competitors had more complete and explicit warnings on their products, which persuaded the Supreme Court of Canada that Lastoplex's own warning was insufficient.

In some cases, an insufficient warning or failure to warn in negligence can rise to the level of the intentional tort of fraudulent misrepresentation based on an analysis of the conduct of the defendant (Osborne, 2015). In a failure to warn in negligence cases, a manufacturer may provide insufficient warning of the risks associated with a product, as in *Lastoplex* (1972) or *Hollis v Dow Corning Corp* [*Hollis*] (1995) discussed in the next section. Conduct amounts to negligence when the conduct itself results in the foreseeable and substantial risk of its consequences (Osborne, 2015). For instance, a manufacturer can be found negligent for design defects, manufacturing defects and failure to warn. Conduct can be elevated from negligence to intentional conduct when the conduct or consequences are substantially certain to result from the manufacturer's conduct (Osborne, 2015). For example, if a manufacturer intends to misrepresent, in order to downplay, unfavourable data on the risks associated with a product, this conduct could be considered to be intentional. This is not to say that the manufacturer intended the consumer to be harmed by the product, but that the manufacturer intended the plaintiff, physician and/or regulators to reasonably rely on the misrepresented data, exposing the plaintiff to undue risk. The motivation for such misrepresentation may be to show more favourable data outcomes and side effects profiles, favourably affecting product sales and profitability for the company from sales of the product.

The tort of negligence is dynamic and often moulded by social norms and policies, characteristically focusing on the quality of the defendant's conduct. A tort of negligence interrogation requires broad analyses informed by the principles of reasonableness and foreseeability (Osborne, 2015). Alternatively, intentional torts analyses tend to be much narrower in scope and are restricted to conservative fact-based analyses. Intentional tort analyses tend to emphasize corrective justice in a manner that assigns responsibility for harm between the specifically involved litigants with an eye to accountability for culpable wrongdoing (Osborne, 2015).

In the fraudulent misrepresentation context, it is plausible that a fraudulent misrepresentation of a material fact that is argued to be associated with harm to the injured party could have prevented or delayed the injured party's realization of their right of action until past the expiration of the 2-year limitation period in Ontario, for example (*Limitations Act*, 2002). By its nature, a finding of fraudulent misrepresentation by a court could indicate that the injured party would reasonably have been prevented from knowing or gaining access to the material information until after the limitation period. At this point, in order to delay the expiration of the limitation period, it would be open to the injured party to advance the equitable doctrine of fraudulent concealment.

9.4 Fraudulent Misrepresentation, and the Equitable Doctrine of Fraudulent Concealment in the Context of the Pharmaceutical Industry

Fraudulent misrepresentation claims may be advanced by plaintiffs in pharmaceutical harms cases wherein manufacturers have published misleading or inadequate warnings regarding the safety, efficacy and side effects profiles of their drug products. Due to the nature of drug harm cases where evidence of misrepresentation of adverse events data is revealed, an injured patient-consumer may not discover that he or she has a cause of action until after the statutory limitation period from the date of harm has expired. In these cases, a plaintiff may advance the equitable doctrine of fraudulent concealment for which the plaintiff must establish a special relationship between the plaintiff and defendant. Defining the special relationship is challenging, considering that such a relationship has not been clearly defined in Canadian case law or legislation. Furthermore, there exists limited literature and case law interpreting the criteria that must be satisfied to establish a special relationship.

Fraudulent misrepresentation is based on the intentional tort of deceit. This facet of tort law aims to protect consumers from deceptive practices in the marketplace, based on the moral principle that business interests must not be advanced by intentional misrepresentation about the nature or quality of the property or services offered to customers (Osborne, 2015). The tort of deceit can be applied in cases of fraudulently induced contracts, property damage, as well as personal injury. In each case, four elements must be satisfied to establish the tort of deceit. First, there must be a *misrepresentation of fact* made by the defendant. Second, the defendant's *misrepresentation must constitute fraud*, which requires proof of dishonesty. Dishonesty can be established with proof that the defendant knew that its statement of fact was untrue, or that the defendant recklessly misstated facts. Third, the defendant must *intend or desire for the plaintiff to rely on its misstatement*, coupled with a*ctual reliance on the misstatement* by the plaintiff. Misrepresentation does not require the statement to be made directly to the plaintiff but requires proof that the defendant desired or intended for the plaintiff to rely on its statement, or the defendant could, with reasonable certainty, have known that the plaintiff would rely on its statement (*Peek v Gurney*, 1873; Osborne, 2015). This requirement of reliance by the plaintiff on the defendant's statement aims to address the element of causation between the fraudulent misrepresentation by the defendant and the resultant loss suffered by the plaintiff. Therefore, there can be no liability on the defendant where the plaintiff knows that the misrepresentation made by the defendant is untrue, or where the plaintiff decides to rely on his or her own judgment or inquiries of the misrepresentation. Fourth, as a result of the plaintiff's reliance on the defendant's misrepresentation, the *plaintiff must have suffered damage*.

If the plaintiff can satisfy the court of these four criteria, the court may award damages. The remedy of damages awarded to the plaintiff is intended not to put the plaintiff in the position he or she would have been in if the representation was true, but, instead, to put the plaintiff in the position that he or she would have been in if the misrepresentation had not been made in the first place. Importantly, liability for misrepresentation of facts within one's possession may not arise unless there exist special circumstances such as either a fiduciary or special relationship between the parties (Osborne, 2015).

In Canadian tort law, it has been well established that manufacturers of products have a duty in tort to warn customers of dangers and risks inherent in the use of their products, about which they knew or ought to have known both at the time of sale and during its time on the market (*Hollis*, 1995). According to the principle of the duty to warn as set out in *Hollis* (1995), manufacturers must reasonably communicate and clearly describe any specific dangers that arise from ordinary use of their products. Furthermore, manufacturers of medical products may shoulder an inherently heavier burden of the duty to warn than other nonmedical manufacturers:

> *Medical products are often designed for bodily ingestion or implantation, and the risks created by their improper use are obviously substantial. The courts in [Canada] have long recognized that manufacturers of products that are ingested, consumed, or otherwise placed in the body, and thereby have a great capacity to cause injury to consumers, are subject to a correspondingly high standard of care under the law of negligence. [...] Given the intimate relationship between medical products and the consumer's body, and the resulting risk created to the consumer, there will almost always be a heavy onus on manufacturers of medical products to provide clear, complete and current information concerning the dangers inherent in the ordinary use of their product* (*Hollis*, 1995).

A successful application of the doctrine of fraudulent concealment prevents a limitation period from being utilized by the defendant as an "instrument of injustice" and will suspend the limitations clock until the point in time at which a reasonable person could reasonably have discovered the cause of action, taking into consideration the extent and nature of the misrepresentation (*Giroux Estate v Trillium Health Centre* [*Giroux Estate*], 2005). The doctrine of fraudulent concealment prevents defendants who are engaged in a special relationship with the plaintiff from using a limitation period as an "instrument of fraud" against the plaintiff when the plaintiff could not have discovered that he or she had a cause of action within that time period due to the misrepresentation or concealment (*Giroux Estate*, 2005; *Roulston v McKenny* [*Roulston*], 2016). Both of these cases deal with the limitation period under the *Trustee Act*, which sets out a 2-year period for actions by trustees of the estate of a deceased person, and which contains no provisions about reasonable discoverability by the plaintiff, unlike the Ontario *Limitations Act* (2002), which incorporates the discoverability rule. For the purposes of this chapter, the *Giroux Estate* and *Roulston* analyses of the special relationship in order to define the relationship is considered.

Fraudulent Concealment and the Element of the "Special Relationship"

The doctrine of fraudulent concealment is to be applied via three elements outlined in *Giroux Estate* (2005). The burden of proving these elements rests on the plaintiff as the party advancing the doctrine. The plaintiff must first establish, on the facts, that the defendant and plaintiff engaged in a special relationship with each other. If the plaintiff can establish the existence of a special relationship, then the plaintiff can move on to establishing that the conduct of the defendant was unconscionable. Last, the plaintiff must establish that the defendant's conduct concealed the plaintiff's right of action. This chapter focuses on the first element of the fraudulent concealment test, establishing the special relationship, the onus of which is borne by the plaintiff.

9.5 Defining the Special Relationship in Canadian Case Law

Although the special relationship has not been clearly defined, it has been referred to in decisions across areas of law. A distilled summary of the various conceptions of the special relationship in the Supreme Court of Canada decisions is comparatively reflected in Table 9.1.

In *Crocker v Sundance Northwest Resorts Ltd.* [*Crocker*] (1988), the Supreme Court of Canada found that where a special relationship can be established between two parties, a duty to act attaches. In *Crocker*, an inviter–invitee situation, the appellant attended a tubing competition, which was hosted by the respondent ski-hill. Prior to participating in the tubing race, the appellant became intoxicated in the ski-hill's bar. Although the appellant was visibly intoxicated at the time of the race, and despite that the race's organizers suggested that he did not participate, they permitted him to participate anyway, resulting in his tubing accident from which he was rendered paraplegic. The court concluded that the relationship between the appellant and the respondent ski-hill gave rise to a positive duty to act, given that the ski-hill hosted an inherently dangerous competition to promote its resort and profitability, provided liquor to the appellant prior to the race, and its employees were aware that intoxication would heighten the chance of injury to the appellant; therefore, the respondent ski-hill should have accepted responsibility for taking all reasonable steps to prevent the appellant's injury.

In *Stewart v Pettie* [*Stewart*] (1995), the Supreme Court of Canada defined the special relationship in the context of an inviter–invitee relationship, where the special relationship involves a positive duty to take steps to discharge harms that are reasonably foreseeable as a result of the impugned conduct. *Stewart* is a negligence case in which a question was raised as to whether a duty of care exists between alcohol-serving establishments and their patrons whose intoxication prevents them from taking to care so as not to harm themselves or other third parties. A duty of care was found to exist between the alcohol-serving establishment, as the inviters, and its patrons, as the invitees, but not between the inviters and the broader motoring public. The inviter–invitee relationship was identified as a special relationship, but the court stopped short of imposing on the inviter the duty to enquire about who was driving and who was not, given that the commercial host knew that some of its party guests remained sober and reasonably assumed that one of the sober guests could, instead, drive.

The court in *Crocker* (1988) stated that "Canadian courts have become increasingly willing to expand the number and kind of special relationships to which a

Table 9.1 Elements of the "Special Relationship" as defined by the Supreme Court of Canada

Stewart v Pettie	Crocker v Sundance	McInerney v MacDonald	Norberg v Wynrib	Queen v Cognos	Imperial Tobacco
Inviter–invitee relationship	Inviter–invitee relationship	Doctor–patient relationship, right of access to medical records in confidence	Doctor–patient relationship, sexual misconduct, consent	Negligent misrepresentation, employer–prospective employee	Tobacco industry–consumers/ government
• Duty of care • Positive duty to take steps to discharge harms that are reasonably foreseeable as a result of the impugned conduct	• Positive duty to act • Responsibility to take all reasonable steps to prevent injury to the plaintiff	• Duty to act with utmost good faith and loyalty • Duty to hold patients' information • Duty to make proper disclosure of information to the patient • Duty to grant access to information used in treatment decisions • Duty to grant access to the records if access would not harm the patient's beneficial interest	• Duty to act with good faith and loyalty • Duty to avoid a conflict of duty and self-interest • Unequal distribution of power • Situational weakness and vulnerability in which trust and confidence has been reposed in the other party	• Interviewer owed a duty of care to the interviewee • Duty of exercising reasonable care and diligence in making representations • Duty to make accurate and not misleading representations • Not a duty to make full disclosure, but disclosure of highly relevant information about the nature and existence of the opportunity • Foreseeable and reasonable reliance cn the representation • Voluntary assumption of responsibility by the representor	• Proximity to the plaintiff • Duty to take reasonable care to prevent a causally related foreseeable loss or harm to the plaintiff • *Prima facie* duty of care • Reasonable foreseeability that a consumer would rely on the company's representations of its product(s) • The company's representations of its product(s) would be reasonably relied upon by consumers

positive duty to act attaches". The court in *Crocker* at paragraph 18 continues to explain this willingness by quoting Allen Linden's 1982 *Canadian Tort Law*:

> *There is a growing group of special relations that import an obligation to engage in positive conduct for the benefit of another. Normally, there is some element of control of some economic benefit inuring to the person as a result of the relation, which justifies the creation of duty. For example, if there is a contract or a bailment, a failure to act may be actionable. It is not enough, however, if the contract is with a third person, as where a doctor agreed with the husband to attend his wife at childbirth. Carriers, innkeepers, warehousemen and public utilities, who hold themselves out to the public as being prepared to give service, are subject to this responsibility. So too, a master may be obliged to provide aid to one of his servants in peril, a shopkeeper to his invitee, a school to a pupil and a shipmaster to a passenger. Obligations to take positive action are also imposed upon occupiers of premises to make their property safe for the reception of certain entrants and for passersby on the highway. A policeman may owe a civil duty to report dangerous road conditions. Institutions which have custody over people, such as hospitals, jails and the like, may be obliged to take reasonable steps to protect those under their care. There will undoubtedly be additions to this list of special situations in the years ahead.* (Crocker, 1988)

Although this quote was published almost four decades ago, the Canadian courts have continued to apply its sentiment of incremental inclusion of special relationships through thoughtful and cautious analysis. Since then, the Supreme Court of Canada has defined the doctor–patient, banker–client, architect–client, engineer–client, solicitor–client, employer–employee (*Queen v Cognos Inc.* [*Queen*], 1993), parent–child (*M(K) v M(H)* [*M(K)*], 1992), First Nations groups-Crown (*Guerin v R* [*Guerin*], 1984), solicitor–client, principal–agent, trustee–beneficiary, director–corporation (*Lac Minerals Inc. v International Corona Resources Ltd.* [*Lac Minerals*], 1989), and bailor–bailee (*BG Checo International Ltd. v British Columbia Hydro and Power Authority* [*BC Hydro*], 1993) relationships as special relationships.

In 1992, the Supreme Court of Canada decided the *McInerney v MacDonald* [*McInerney*] (1992) and *Norberg v Wynrib* [*Norberg*] (1992) cases, in which the court advanced the notion that the physician–patient relationship amounted to one that is fiduciary in nature, resulting in certain duties arising from the special relationship that ensues due to the elements of trust and confidence which are inherent to the relationship. In *McInerney*, the court defined the duties arising from the special relationship as well as the fiduciary relationship to include: (i) the duty of the physician to act with utmost good faith and loyalty, (ii) to hold patients' information in confidence, (iii) to make proper disclosure of information to the patient, (iv) the obligation to grant access to the information used in administering treatment, (v) to grant access to the patients' records, within limits to be decided by the physician in the best interest of the patient.

In *Norberg*, the doctor–patient relationship was defined in the context of sexual misconduct allegations with an explicit awareness of the power imbalance between physicians and their patients. This power imbalance was expressed by describing "a special 'power dependency' relationship", clarifying that the existence of a power imbalance does not itself necessarily determine the existence of a special relationship and that a factually specific analysis must be undertaken for each case in which a special relationship may exist. Since the special relationship between a doctor and patient follows from the fiduciary relationship that defines the relationship, the court provides three elements of a fiduciary relationship: (i) the fiduciary has scope for the exercise of some discretion or power, (ii) the fiduciary can unilaterally exercise that power or discretion so as to affect the beneficiary's legal or practical interests, and (iii) the beneficiary is peculiarly vulnerable or at the mercy of the fiduciary holding the discretion or power. Furthermore, the court states that physicians owe their patients the quintessential duties associated with a fiduciary relationship, namely, the duty to act with loyalty, good faith and avoidance of conflict of duty and self-interest.

In the context of physician–patient relationships, the characteristics of the fiduciary duty and special relationship may ostensibly overlap; however, with some careful analysis, the special relationship elements may be parsed. According to *Norberg*, a physician's duty within the doctor–patient relationship is based in contract and can be subject to action in either contract where there is an overwhelming power imbalance between the parties or tort in negligence. The court goes on to explain that "…fiduciary duties should not be superimposed on common law duties" (*Norberg*, 1992), so as to attempt to differentiate the fiduciary relationship from the special relationship. A comparative situational weakness between the parties seems to be an important analytical consideration in this case, where the "weakness", which is to be factually established in each case, "…may arise out of a special relationship in which trust and confidence has been reposed in the other party" (*Norberg*, 1992). In the physician–patient relationship, this comparative weakness, resulting in a power dependency relationship, is underscored by the professional association of the physician and the physician's position of gatekeeper to medical treatments.

Situational vulnerability is an additionally important element to consider in analyses of special relationships. In *Norberg*, although the issue on the facts was sexual misconduct between a doctor and patient, the notion of vulnerability as a patient in a medical situation is applicable across issues in various medical milieus. In *Norberg*, the appellant was found to be in a vulnerable position as a result of drug dependence, placing the appellant in a situation in which making an unconstrained and uninfluenced choice was diminished. In the physician–patient relationship, the fiduciary nature of the relationship is recognized by law owing to the material discrepancy in the circumstances of the relationship placing the physician in the position of power and the patient in the position of vulnerability. Without this material discrepancy, the court explains, the parties would be on relatively equal footing thereby allowing a contract and tort analysis to be appropriate. The addition of the element of vulnerability in the relationship elevates the physician–patient relationship to the fiduciary level.

Special relationships have also been described in the employer–interviewee context in the Supreme Court of Canada decision of *Queen* (1993), where the interviewer made misrepresentations to the interviewee about important elements of the position. In this case, the court defined the test for the tort of negligent misrepresentation, which is comprised of five criteria that must be established: (i) there must be a duty of care based on a special relationship, (ii) the representation in question must be untrue, inaccurate, or misleading, (iii) the representor must have acted negligently in making the misrepresentation, (iv) the representee must have relied, in a reasonable manner, on the negligent misrepresentation and (v) the reliance must have been detrimental to the representee in the sense that damages resulted (*Queen*, 1993). Importantly, this decision makes clear that actions may be advanced even in situations in which the negligent misrepresentation occurs in a pre-contractual setting, such as a job interview, although the employment contract may have relevant provisions that prevent the action in tort, leaving only remedies in contract law. The court also found that the interviewer owed a duty of care and to exercise reasonable care and diligence in making representations concerning the employer and the employment opportunity being offered. Accordingly, the interviewer owed a duty of care to both the employer and the interviewee to the objective standard of the reasonable person making such representations that are accurate and not misleading. A special relationship was found to result from the relationship between the interviewer and interviewee, giving rise to a duty of care.

The court in *Queen* (1993) states that there is an academic debate about the proper test that should be applied to ascertain when a special relationship exists between a party representing facts and the receiver of those representations, but that it is unnecessary for the court to engage in this debate. Despite this assertion, the court acknowledges the perspectives that a "foreseeable and reasonable reliance" on the representations is essential to a special relationships analysis, while also acknowledging the element of a "voluntary assumption of responsibility" by the party making the representations. The court subsequently goes on to apply its own criteria to define the special relationship as between the representor and the interviewee: (i) representor assumes responsibility for the representations, (ii) foreseeability by the representor that the interviewer would detrimentally rely on the representations, resulting in damages to the interviewer, (iii) existence of a relationship of proximity between the parties at all material times, (iv) existence of a duty of care between the representor and the interviewee; and (vi) duty of care and honesty (*Queen*, 1993). The court continues that the profession of the representor may on the facts provide sufficient indications of a duty of care; however, profession alone is not a determining factor, supporting instead a "...more flexible approach" to determining the special relationship.

M(K) was a (1992) Supreme Court of Canada case in which the parent–child relationship was interrogated on the basis of sexual assault and incest occurring between a father and a daughter. Although this case centres on the fiduciary relationship between parent and child, this case mentions the special relationship. The court in *M(K)* cites *Kitchen v Royal Air Forces Association* (1958) for its conceptualization of the special relationship as being an important element of a situation in which the doctrine of equitable fraud can be applied and against which a person can be protected. The element of unconscionability is also described as an essential element of the application of the doctrine of equitable fraud, and so is also somehow associated with the notion of the special relationship. The decision moves then to unpacking the fiduciary relationship between a parent and child, some elements of which have been reflected in conceptualizations of the special relationship in other cases.

Guerin was a (1984) Supreme Court of Canada decision, which dealt with the nature of the relationship between the Crown and First Nations groups in Canada, ultimately deeming this relationship to be fiduciary in nature. The facts of the case were that the Musqueam First Nations band in Canada held a large acreage of favourable land in the Vancouver area. The Canadian federal government made a deal with a golf club on behalf of the band, without accurately expressing to the band the conditions of the agreement between the government and the club. As in *M(K)*, the court cites *Kitchen v Royal Air Force Association* (1958) for its use of the term special relationship as being associated with equitable fraud and unconscionable conduct between the parties.

Finally, *Imperial Tobacco* was a 2011 Supreme Court of Canada decision concerning two appeals, first, the *Costs Recovery* case in which the Government of British Columbia sought to recover costs for paying for the medical treatment of individuals suffering from tobacco-related illnesses and, second, the *Knight* case, which was a class action. The class action was brought against Imperial Tobacco Company Ltd. on behalf of class members who purchased "light" or "mild" cigarettes that the class alleged contained higher levels of toxins than had been advertised on the cigarette packages, further alleging that the toxin levels in the subject cigarettes were just as harmful as toxin levels in regular cigarettes. These allegations were founded in the claims of negligent misrepresentation, negligent design and failure to warn.

The special relationships question raised in *Imperial Tobacco* is one within negligent misrepresentation, where the court stated that a special relationship is formed between the parties if, on the facts, there is a *prima facie* relationship of proximity in which a failure to take reasonable care might foreseeably cause a loss or harm to the plaintiff. The court goes on to state that both of these requirements for a *prima facie* duty of care in a negligent misrepresentation case are considered to be established when a special relationship exists between the parties. In perhaps the most clearly defined analysis in the case law for establishing a special relationship, the court in *Imperial Tobacco* states that a special relationship will be considered to be established where (i) the defendant ought reasonably to foresee that the plaintiff will rely on its representation, and (ii) that the defendant's representation would reasonably be relied upon by the plaintiff in the circumstances. The court continues to provide a second stage of the analysis, which assesses whether this *prima facie* duty of care ought not to be recognized for policy reasons.

The common thread sewn through the various conceptualizations of the special relationship across areas of law is the duty to not place another person into a position in which it is foreseeable that they might suffer injury. The themes of trust and confidence in the parties in positions of power, as well as their disclosure of information for the beneficial interest of the plaintiff is also interwoven into these cases.

The Special Relationship Element of the Equitable Doctrine of Fraudulent Concealment: GlaxoSmithKline, Paxil and its Patient-Consumers

When a plaintiff advances the doctrine of fraudulent concealment and proceeds in the endeavour of establishing a special relationship between himself or herself and the defendant manufacturer, prior to being able to satisfy the second and third elements of the fraudulent concealment test, the plaintiff will likely be faced with an uphill battle. The defendant manufacturer will likely advance the following positions: that there exists no special relationship between the two parties due to the learned intermediary doctrine, that the relational distance between the two parties effectively prevents a special relationship from manifesting between the plaintiff and defendant, and that although there may be a relationship between the manufacturer and consumer, the relationship does not amount to a fiduciary relationship between the two parties.

Since neither the special relationship nor its constituting criteria have been definitively established in law, it is possible that such a relationship could be argued on the facts given the unique business structure of the pharmaceutical industry and the role of Health Canada and physicians in the pharmaceutical market. An analysis of specific steps in the pharmaceutical business structure, in combination with agency theory principles in the context of pharmaceutical industry practices, may provide some insights into the possibility of establishing a special relationship between a drug company and its ultimate consumer. Establishing the existence of this relationship could result in a judge deciding to delay, in equity, the limitation period where expiration of the limitation period has become germane to the proceedings, or a finding of negligence in a failure to warn case.

9.6 Interrogating the Relationships Between the Manufacturer and Consumer in the Context of the Pharmaceutical Industry's Business Structure Pertaining to Clinical Trial Research and Academic Publishing

The business structure of the pharmaceutical industry is multilayered and multifaceted (See Gagnon, 2009; Mirowski, 2011). An analysis of relationships within the pharmaceutical industry requires consideration of the structure and context in which data-centred decisions are made and to whom a duty is owed. The service and product supply chain of the pharmaceutical industry is similarly unique and important to consider in any analysis concerning the relationships between drug companies and their ultimate consumers. The landscape of science and business has undergone notable transformation since the 1980s. Since this time, the pharmaceutical sector worldwide has experienced tremendous growth, which has coincided with shifts in the nature and structure of its research to become increasingly privatized and driven by the commodification and commercialization of knowledge, rather than funded publicly (Mirowski, 2011).

The research, development and promotion of pharmaceuticals has been influenced and shaped by converging social, economic and political circumstances that have manifested in increasing deregulation of industry and its function in the free market, as well as outsourcing some of its core roles, including research (Enyinda, Briggs, &

Bachkar, 2009; Mirowski, 2011). The shift in the ways in which knowledge is considered and used in the process of privatized pharmaceutical development has brought with it a unique supply chain, which some would argue positions drug companies and their research at greater distances from their ultimate consumers. However, the control that drug companies maintain over the release of their data arguably positions the companies in the closer relational distance to their ultimate consumers, since the ultimate consumer must rely on the data disclosures by the company. This reliance on the company by the consumer intensifies when the outsourcing of clinical trials, data collection, analyses and interpretation is considered.

Supply chain management is central to the coordination of services and production of goods, which, ultimately, are sold to the consumer at a profit (Hugos, 2011). The individual elements of a supply chain often have conflicting needs, for instance, producing enough product while maintaining efficiency, which calls for the reduction of inventory (Hugos, 2011). While it can be more straightforward identifying risks to consumers in the manufacturing and distribution phases of the supply chain, which may manifest as negligence claims for contamination, it is challenging and, sometimes, almost impossible to identify risks to consumers that originate from the research and communication phases of the pharmaceutical business structure and supply chain. Therefore, this analysis is concerned with the initial stages of the pharmaceutical supply chain, namely, the research and data communication processes, which are crucial to the grooming and preparation of the market for the anticipated approval of a company's pharmaceutical product.

Drug companies understand that the pharmaceutical supply chain is extremely complex, requiring long-term planning and market engagement, beginning even before a drug is approved for the market. The shift from medical scientific research being conducted in academic medical centres to being privatized through academic–industry partnerships has required a new management structure, finely tuned to the rhythm and needs of drug companies throughout the R&D and communication processes. Increasingly, drug companies are globally outsourcing their research to companies called contract research organizations (CROs) and their medical writing and communications to companies called medical writing organizations (MWOs) and medical communications companies (MCCs) (Mirowski, 2001).

9.7 The Learned Intermediary and Material Representation

Generally, manufacturers' duty to warn is owed directly to the ultimate consumer. In exceptional situations, however, the defendant may argue that its duty to warn is satisfied through its application of the learned intermediary doctrine, a common law principle which allows for a fact-based analysis to determine whether intermediate examination and intervening cause releases the manufacturer from its duty to warn directly to the consumer (*Donoghue*, 1932). The learned intermediary doctrine is applicable in two main circumstances: first, where the nature of the manufacturer's product is highly technical and the manufacturer intends the product to be used only under expert supervision and, second, where it is unlikely or unrealistic for the consumer to receive a direct warning from the manufacturer prior to use of the product (*Hollis*, 1995).

The Supreme Court of Canada reinforced the *Lambert* principle in *Hollis* (1995), a case in which the plaintiff was injured as a result of a silicone breast implant rupture. The manufacturer knew of the small risk of rupture with regular activity but failed to warn her or her physician of the risk. In *Hollis* at paragraphs 21 and 22, the court held that:

> When manufacturers place products into the flow of commerce, they create a relationship of reliance with consumers, who have far less knowledge than the manufacturers concerning the dangers inherent in the use of the products, and are therefore put at risk if the product is not safe. The duty to warn serves to correct the knowledge imbalance between manufacturers and consumers by alerting consumers to any dangers and allowing them to make informed decisions concerning the safe use of the product.
>
> The nature and scope of the manufacturer's duty to warn varies with the level of danger entailed by the ordinary use of the product. Where significant dangers are entailed by the ordinary use of the product, it will rarely be sufficient for manufacturers to give general warnings concerning those dangers: The warnings must be sufficiently detailed to give the consumer a full indication of each of the special dangers arising from the use of the product. (paras 21 and 22)

A key fact in *Hollis* was that the physician was accepted by the court to function as the learned intermediary; therefore, the medical device manufacturer ought to have warned the physician of the risk of rupture without warning. The court further states that a breast implant is distinguishable from most other manufactured goods since the implant and its packaging are not placed into the hands of the patient-consumer. Rather, the implant is obtained from the manufacturer by the surgeon, who is in the best position to read any safety warnings provided by the manufacturer (*Hollis*, 1995). The court states that, in this way, breast implants can be viewed as being "…analogous to prescription drugs, where the patient place primary reliance for information on the judgement of the surgeon, who is a "learned intermediary", and not on the manufacturer" (*Hollis*, 1995). The court distinguishes breast implants from medications such as oral contraceptives, which are widely prescribed, widely used and unlike breast implants upon their implantation, do not require the intervention of a physician at each time that the medication is ingested. There are several parallels, however, between the medical devices and drug products, since pharmaceutical products are both studied and manufactured by the sponsoring company and the consumer must obtain the product from the company, through a prescription from a physician and filling the prescription through a pharmacist.

Application of the learned intermediary doctrine does not absolve the manufacturer of liability. The learned intermediary doctrine requires the intermediary to be fully apprised of the risks and can only be argued successfully by the defendant when the defendant manufacturer has taken adequate steps to ensure that the intermediary's knowledge of the risks associated with the drug or device product approximates that which is possessed by the manufacturer. In the drug safety world of prescription medicines, there are intermediaries such as Health Canada as the regulator and physicians as gatekeepers to prescriptions; however, despite the presence of these intermediaries, consumers who receive prescription medications are required to receive and have access to product labels and monographs, which are documents that describe the risks for adverse events associated with manufacturers' prescription medications, among other content. Although Health Canada and physicians would likely be considered to be in the most effective positions as intermediaries, the question of whether regulators and physicians are learned intermediaries on the facts remains. In the Paxil case study, neither Health Canada nor physicians would have had access to the raw adverse events data from Study 329. In addition to the misrepresented adverse events data from Study 329, several clinical trials that had unfavourable conclusions regarding the safety and effectiveness of Paxil were suppressed and remained unpublished, skewing the available data on Paxil (Bass, 2008).

9.8 The Role of the Medical Professional: Is the Physician a Learned Intermediary in the Special Relationship Analysis?

Physicians globally tend to receive their information on pharmaceutical products from a few sources: medical school, medical journal articles, clinical practice guidelines, drug company sales representatives and continuing medical education (CME) or continuing professional development (CPD).

9.8.1 Informing Physicians: The Medical Ghostwriting of Study 329 on Paroxetine (Paxil)

The U.S. lawsuit, *Beverly Smith v SmithKline Beecham*, revealed internal industry documents, providing evidence that Study 329 was ghostwritten by a communications agency called "Scientific Therapeutics Information, Inc." (STI). Industry-sponsored ghostwriting is common across heavily marketed drug classes and companies that sponsor the manufacture and promotion of these classes. Medical ghostwriting has been a subject of scholarly debate. While some argue that medical ghostwriting is not an ethically compromised practice, others have provided evidence that medical ghostwriting specifically operates in a manner that conceals the true involvement of the sponsoring drug company through its relationships with the contract research and writing organizations that work to foster favourable research results for the sponsoring company, and its financial conflict of interest relationships with physicians and researchers who promote the benefits of the pharmaceutical products in the published medical literature.

MCCs, like STI, compete with bids for their management of the publication process when developing a manuscript for the favourable profiling of a drug in the medical literature on behalf of the sponsoring company in a tendering-like process (Laden, 1998). STI offered for $17,250 to "provide all necessary resources to complete this manuscript including writing, editing, library research and retrieval, copy editing, proof reading, word processing, art work, and the needed coordination with author(s), sponsor, and journal." It was not until the fourth version of the Study 329 manuscript that STI developed a plan to send the manuscript to all authors "for comment" only. In depositions, the associate editorial director of STI, Sally Laden, stated that she was involved in interpreting the clinical trial data and her ideas generated the manuscript, without assistance from the named guest authors on the final manuscript. In her testimony, Laden claimed that she coordinated the publications process, which included responding to peer-reviews, the journal editor's comments, and providing submission packages for the lead guest author, Keller, including draft cover letters for the editors of the prospective journals. SmithKline Beecham and STI maintained careful control of the data, prohibiting any publication of the data without SmithKline Beecham's approval, as the data was proprietary to the company. From the published Study 329, we see that the guest authors included the side effect of "emotional lability" and not "suicidality". Presumably, if the named authors, as physicians, knew that "emotional lability" was a coding term that was used to mask "suicidality", they might have shown more concern for the higher rates of "emotional lability" on paroxetine.

By the end of the research collection and analysis processes of the pharmaceutical supply chain, the fate of the subject drug's side effects profile is sealed, until potential future litigation calls for documents disclosure, as was the case with paroxetine. Because a drug is not publicly regulated until it is approved for the market, and companies conduct and regulate their own clinical trials and analyses, there is no opportunity for intermediate examination. Visual inspection of pharmaceuticals cannot detect fraudulent misrepresentation of clinical trial data. The relevant type of intermediary inspection that would be beneficial to medication consumers would be independent reviews of the primary patient-level data by academic researchers early on in the research level of the supply chain. No inspection of the manufacturing, distribution or retail processes could identify the suppression of clinical trial data, where fundamental characteristics of a drug's safety profile have been hidden, as was the case with paroxetine.

Drug companies' outsourcing of participant recruitment, conducting clinical trials and the collection of data, which are essential elements of the pharmaceutical supply chain, has developed alongside the development of MCCs. MCCs are another relatively new entity in the pharmaceutical industry's supply chain. Like CROs, MCCs have grown into their own industry of "publication planners", as they are termed in the pharmaceutical promotion scholarship. These professionals are now supported and represented by two international associations, the International Publication Planning Association and the International Society for Medical Publication

Professionals (ISMPP), which, alone, has 1400 members (https://www.ismpp.org/initiatives). Publication planners, including medical writers, work with the sponsoring drug company to shape the publication content and control the publication cycle. Controlling the publication cycle and manuscript contents has become an important role that MCCs offer because drug companies view research as branding, resulting in brand recognition and loyalty of physicians and patient-consumers (Smith, 2005). In the pharmaceutical industry, published research that results in favourable conclusions is essential marketing tools (Smith, 2005). A favourable clinical trial that is published in a prominent peer-reviewed medical journal can be worth thousands of pages in advertising, especially since the conclusions can subsequently be endorsed in press releases from both the medical journal and the sponsoring drug company's retained public relations or communications firm (Smith, 2005). These published studies also become the focus and topics of continuing medical education (CME) and continuing professional development (CPD) content for physicians.

The use of contract research, writing and communications to these entities, which work with the sponsoring company to control and shape the multiple steps in the processes of R&D, clinical trials and clinical trial management, data collection and analyses, regulatory submissions, as well as drafting (ghostwriting) of manuscripts that later appear in the medical literature under the name of another physician or academic medical researcher (McHenry & Jureidini, 2008; Sismondo, 2007). This multi-year pharmaceutical business structure and supply chain as been termed "ghost management" because the role of the drug companies and their contracted entities in shaping the narrative and discourse on their drugs are unknown to physicians and the public. This management of every step in a drug's lifespan, from pre-approval through to patent-expiration, is effectively masked by financial conflict of interest relationships with academic physicians and researchers who, for example, appear as "guest" authors on academic peer-reviewed medical journal articles, which had previously been ghostwritten by medical writers at MWOs (Fugh-Berman & Dodgson, 2008; Sismondo, 2007).

9.8.2 Informing Physicians: Industry-Sponsored and-Conducted Data Analyses

The evidence suggests that GSK's own researchers were responsible for the data interpretation decisions made in Study 329; however, drug company researchers operate similarly to CRO researchers by overseeing the clinical trial at its various locations, undertaking analyses and interpretations of data, and preparing the data for the manuscript writers (Mirowski, 2011). In Study 329, the participants' side effects outcomes during the clinical trial were coded using a coding "dictionary", which is standard practice; however, the dictionary that is used matters and can alter the interpretations and meaning of the data. For instance, the ADECS dictionary that was used in the Study 329 side effects analysis failed to accurately represent the serious adverse events experienced by the participants and skewed the preliminary data. Furthermore, when only the analysed data tables are presented to the manuscript writers and, then, eventual guest authors of a study, it is impossible for them to understand the serious risks of harm associated with the drug. Typically, it is only these tables and not the patient-level data from the clinical trials that the MCCs and eventual guest authors receive.

A review of the literature has revealed that some scholars have expressed concerns about the potential for undue influence on research integrity when clinical studies are industry funded. Financial relationships between academic researchers and industry in the fields of biotechnology and biomedical research have faced criticism. Academic researchers who participate in these financial relationships engage in FCOI relationships, which are likely to be expressed in their research. For example, when research is industry sponsored, there is an increase in the likelihood that results that are favourable to industry sponsors are systematically selected for publication, while results that are deemed to be unfavourable to sponsors are suppressed (Bekelman, Li, & Gross, 2003).

Darren Zinner and colleagues (2009) found, in a sample of 2,168 faculty members from clinical (n = 1,071) and nonclinical (n = 1,097) departments, that some faculty who received industry support for their research experienced restrictions on their permitted communications about their research as well as their choice of research. In this study, 12.9 percent of faculty produced research that resulted in trade secrets, or information that was to be kept confidential to protect its proprietary value. Zinner and colleagues (2009) also found that when faculty were able to choose their research topics, their choices were affected either somewhat or greatly by the prospect of their results having a commercial application. Faculty with industry funding were also more likely than faculty without industry funding to report the delaying of a publication for six or more months, or that the delay of a publication was intended to prevent the publication of unfavourable results (Zinner et al., 2009). These findings are supported and complimented by an earlier study by David Blumenthal and colleagues (1996), which found similar consequences for research transparency among faculty in 1996, indicating that these problems persisted despite advances in FCOI policies.

Products tend to be systematically favoured in published research that is sponsored by the drug manufacturer (Lexchin, Bero, Djulbegovic, & Clark, 2003). A Cochrane Collaboration review of 48 studies by Andreas Lundh and colleagues (2012) found that both pharmaceutical and medical device industry-sponsored studies reported more favourable efficacy results, harms results, and overall conclusions than did studies that were funded by other sources (Lundh et al., 2012). Lundh and colleagues (2012) found that in 14 papers, which included 1,588 drug studies, industry-sponsored studies had favourable efficacy results with significant P-values more often than did non-industry-funded studies (Lundh et al., 2012).

This Cochrane review also found that two papers, which included 131 industry-funded clinical trials on statins and thiazolidinediones, reported that when trials compared two drugs made by competing companies and each company sponsored a trial, the superior drug was reported to be the one made by the sponsoring company (Lundh et al., 2012). An additional paper, which included 20 selective serotonin reuptake inhibitor (SSRI) head-to-head trials, found that all 20 (100%) favoured the treatment that was manufactured by the sponsor of the trial, while none reported results that favoured the comparator drug. These findings of the relationship between sponsorship and study results are duplicated in the findings in the Cochrane report about trial conclusions. In 24 papers, which included 4,616 studies of which 4,403 were drug studies, industry-sponsored studies were more likely to reach conclusions favourable to the sponsor's product (Lundh et al., 2012).

Overall, Lundh and colleagues (2012) found that industry-funded studies obtained results and conclusions that were more often favourable to the sponsors' products than when studies were funded independently from the industry. This review also found that when studies were industry sponsored, there were discrepancies between reported results and the wording of the conclusions, as compared with non-industry funded studies (Lundh et al., 2012). In practice, this means that the conclusions tended to overstate the benefits presented by the results. Industry-funded studies were found to be just as methodologically rigorous as non-industry-funded studies when the traditional methods of measuring rigour, i.e. use of randomized controlled trials (RCTs) or blinding, were evaluated. However, evaluating only these traditional indicators of study quality may lead to missing the more subtle techniques that bias trials and lead to pro-industry results (Lundh et al., 2012).

Execution of clinical trials according to a priori protocols, as well as the objective interpretation and publication of results, can also be influenced through agreements and contracts that stipulate that the sponsoring drug company has access to trial data or that give it the power to restrict or prevent publication of the results (Schott et al., 2010). There are a number of different stages and characteristics of clinical trials that are influenced by drug company sponsorship. For example, internal industry documents have provided evidence that drug companies have failed to reveal relevant data on adverse drug reactions (ADRs) to the public or the FDA at the appropriate times. For example, the manufacturer of cerivastatin was aware of the drug's interaction with gemfibrozil, which leads to an increased occurrence of rhabdomyolysis (muscle breakdown), approximately 100 days after the product was launched on the market. It took 18 months, however, for this interaction to be added to cerivastatin's product information on contraindications. Another example is that the manufacturer of rofecoxib had data on increased mortality in Alzheimer's dementia patients, but failed to communicate this to both patients and the FDA in a timely manner. The manufacturer of rofecoxib also failed to adequately evaluate the occurrence of cardiovascular ADRs (Schott et al., 2010). Physician and FDA official, David Graham, testified on the harms data for rofecoxib and estimated that approximately 100,000 excess cases of heart attack and sudden cardiac death occurred (Graham, 2004).

The manufacturer of paroxetine (Paxil) failed to include known ADRs in the drug's accompanying information. A 2008 study by Ivar Aursnes and Marianne Klemp Gjertsen explored the ADRs that had already been documented at the time that paroxetine was first licensed in 1989 (Aursnes & Klemp Gjertsen, 2008). The authors gained access to the 1989 clinical data on paroxetine that was presented to drug agencies globally (Aursnes, Tvete, Gassemyr, & Natvig, 2005) after an unusual decision by the Norwegian Civil Ombudsman that the Ministry of Health should permit them to examine these documents. Out of 32 ADRs that the test group reported, only 8 were listed as common, despite 19 of the 32 ADRs being statistically significantly more common in the test group compared to the control groups. As of 2008 when this study was published, 5 of these 19 ADRs were still not mentioned in the Summary of Product Characteristics (SPC), which accompanies Patient Information Leaflets (PILs) for licensed medicines in Europe (Aursnes & Klemp Gjertsen, 2008). Included within these five unlisted ADRs were paresthesia and nervousness, both of which were still not mentioned in paroxetine's information in 2010 in Norway, over two decades after the findings were first known (Schott et al., 2010).

A 2015 research study by Joanna Le Noury and colleagues reanalysed SmithKline Beecham's (now, GlaxoSmithKline, GSK) Study 329 on the safety and efficacy of paroxetine compared to imipramine, to determine whether access to and reanalysis of the full data set from an RCT would result in different findings than in the study originally published in 2001 (Le Noury et al., 2015b). Le Noury and colleagues gained access to 77,000 pages of de-identified case report forms (CRFs). They found that neither paroxetine, nor imipramine proved to be effective when it was used to treat major depression in adolescents. In their reanalysis, they also found a clinically significant increase in harms with both imipramine and paroxetine (Le Noury et al., 2015b). These findings are extremely important because they both contrast with the originally reported and interpreted results in the original 2001 publication. Le Noury and colleagues also found that the original study employed biasing techniques in the coding of adverse events in the study, which masked important differences in the suicidal behaviours of the original participants in the study. The decision-making trail, made available by Le Noury and colleagues in Box 2, Appendix 3 (Le Noury et al., 2015a), leads the readers to the unexplainable coding choices made by the original Study 329 team. These coding choices transformed serious adverse events, such as intentionally swallowing 80 Tylenol tablets, into the misrepresentative and completely minimized category of "emotional lability" (Le Noury et al., 2015b).

As the above example shows, techniques used by companies to bias study outcomes are both complex and comprehensive and are utilized at every level of the drug evaluation process, making them exceptionally difficult to detect (Lexchin, 2012). Some additional techniques include enrolling healthy patients in phase III trials, purposefully administering an insufficient dose of the test or comparator drug, and testing the experimental drug against placebo when there are active medication alternatives on the market (Bero, Oostvogel, Bacchetti, & Lee, 2007; Chopra, 2003; Rochon et al., 1994).

9.8.3 *Informing Physicians: Clinical Practice Guidelines (CPGs)*

Clinical practice guidelines (CPGs) play a pivotal role in not only informing physicians about treatment standards, but also in decision-making in physicians' prescribing and treatment choices for their patients. CPGs are widely distributed by professional medical associations and consulted by physicians for the best available clinical evidence; however, the development of CPGs has attracted debate. Despite recommendations that guideline development should be transparent, rigorous and use scientific evidence, clinical experiential knowledge, and patient values to inform and improve recommendations (IOM, 2011), international studies have called into question whether guidelines are developed in this manner. These studies have found that many guidelines have made recommendations based on expert opinion rather than clinical trial data, consensus statements and retrospective case studies rather than data that has integrity and are based on incomplete and inappropriate use of available evidence (Abramson & Starfield, 2005; Cosgrove, Bursztajn, Krimsky, Anaya, & Walker, 2009; Dinnes, Hewison, Altman, & Deeks, 2012; Kung, Miller, & Mackowiak, 2012; Mendelson, Meltzer, Campbell, Caplan, & Kirkpatrick, 2011).

Of note in the context of this chapter is the study by Cosgrove et al. (2009) in which they found that of the authors on clinical practice guidelines from the American Psychiatric Association, 90% had at least one financial conflict of interest relationship with the pharmaceutical industry. The authors also found that in three major psychiatry clinical practice guidelines, 90% of the authors had financial ties to companies that manufactured drugs that were identified in the guidelines as pharmaceutical products that were recommended as a treatment for the subject mental illnesses. Furthermore, concerns about the validity of guideline recommendations have been raised because of the potential for bias during the guideline development process (Bell et al., 2013; Guyatt et al., 2010; Spielmans, & Parry, 2010).

Studies on guidelines in the United States and Europe have demonstrated concern regarding the prevalence, underreporting and consistency of guideline authors' disclosures of their FCOI relationships with the pharmaceutical industry (Brix Bindslev, Schroll, Gotzsche, & Lundh, 2013; Choudhry, Stelfox, & Detsky, 2002; Cosgrove et al., 2009; Langer et al., 2012; Mendelson et al., 2011; Neuman, Korenstein, Ross, & Keyhani, 2011; Norris, Holmer, Ogden, Burda, & Fu, 2013; Papanikolaou et al., 2001). To provide an assessment of the authors' FCOI relationship disclosures on clinical practice guidelines in Canada, we conducted a case study analysis on authors' disclosure statements in 28 guidelines, most recently reviewed or published in 2012 or 2013, drawn from the Canadian Medical Association Infobase. It is reasonable to analyse guidelines reviewed or published during this time because the issue of FCOI disclosures of guideline authors was, and continues to be, a relevant and public issue. We found that, in general, guideline authors commonly disclosed FCOI relationships with industry (Shnier, Lexchin, Romero, & Brown, 2016).

FCOI relationships held by the authors were often with the drug companies that manufactured the medications recommended as first-line treatments in the respective guidelines. We also found that some authors who were on more than one guideline disclosed different FCOI relationships in their statements across guidelines. The findings from this study support the need for additional research to assess the prevalence of FCOI in guidelines in relation to recommended drugs in guidelines. For example, future research which assesses the quality of evidence used in guidelines, where the authors have FCOI relationships and whether the recommendations reflect the evidence would be valuable.

9.9 Informing Physicians: Drug Company Sales Representatives

Studies from countries around the world have shown that 80–95 percent of physicians regularly have interactions with drug company representatives. Physicians engage in these interactions, despite the research that the information provided by drug representatives is overly positive and the result is that physicians' prescribing practices are less appropriate (Moynihan, 2003). In fact, with rare exceptions, studies that have assessed physicians' exposures to information originating from drug companies have found associations with higher prescribing frequency, higher costs and lower prescribing quality (Grande, Frosch, Perkins, & Kahn, 2009; Spurling et al., 2010; Symm, Averitt, Forjuoh, & Preece, 2006).

Most practicing physicians in the United States have relationships with drug companies. A survey by Campbell and colleagues (2010) assessed the difference in the frequency with which physicians in seven medical specialties received payments from the industry in 2004 and 2009. While the frequency with which physicians received payments from industry decreased from 2004 to 2009, physicians having financial relationships with drug companies remained, on the whole, in the majority. The percentage of doctors who received any drug samples, gifts, reimbursements, payments for speakers' bureaus, consulting, advisory boards, and enrolling patients in trials in 2004 was 94 percent and in 2009 was 83.8 percent (Campbell et al., 2010).

Campbell and colleagues (2010) attributed the decrease in physician–industry relationships in the United States to several factors including increased awareness and public attention by the media and professional organizations to issues of FCOI relationships. The decrease in physician engagement in FCOI relationships may also have been due to new policies adopted by some medical schools and hospitals in the United States, which have banned certain types of FCOI relationships including receiving drug samples, industry-sponsored meals and participation in speakers' bureaus. Increased publicly accessible reporting of FCOI relationships in the United States by drug companies, medical schools, states and the federal government may have also contributed to physicians' decreasing engagement in FCOI relationships with drug companies (Campbell et al., 2010).

Research on prescribing behaviours following industry-sponsored symposia accompanied by monetary rewards illustrates that physicians' prescribing practices tend to be skewed to favour the drugs presented. A study by James Orlowski and Leon Wateska (1992) observed physicians' prescribing practices after the physicians attended all-expenses-paid trips to vacation destinations for educational symposia that were sponsored by the manufacturer of the drug being presented. The impact of the trips was evaluated by tracking pharmacy inventory usage reports both before and after the symposia. Both drugs featured in the symposia were relatively new and available only intravenously. Orlowski and Wateska (1992) found that the majority of physicians who attended the symposia insisted that their prescribing behaviours were in no way influenced by accepting elaborate enticements from pharmaceutical companies. Some physicians admitted enticements might make them consider a drug that they might not have otherwise thought of prescribing, while others believed that symposia may convince them that a drug had uses or benefits for their patients that they had not otherwise considered. When the authors studied changes in physicians' prescribing behaviours before and after the symposia, they found that after the symposia, prescribing patterns significantly increased for the two drugs that were the subjects of the symposia. Importantly, these two drugs were no better than the drugs that the doctors were already prescribing (Orlowski & Wateska, 1992).

The finding that accepting items of monetary value from drug companies skews prescribing practices continues to be supported by the literature. James Yeh, Jessica Franklin, Jerry Avorn, Joan Landon and Aaron Kesselheim (2016) conducted a study to determine the association between drug company payments to doctors and prescribing of brand-name compared with generic statins. Yeh and colleagues (2016) found that industry payments are associated with increased rates of physicians prescribing brand-name statins and, for every US$1000 received, prescriptions of brand-name statins increased by 0.1 percent ($P < 0.001$). Similarly, drug company payments for educational training was associated with a 4.8 percent increase in brand-name prescribing ($P = 0.004$).

Another 2016 study by Ryann Grochowski Jones and Charles Ornstein (2016) of ProPublica, a non-profit investigative journalism organization, found that physicians who received larger payments from drug companies more frequently prescribed brand-name drugs as compared with physicians who did not receive payments from drug companies. Grochowski Jones and Ornstein (2016) also found that the type of payment had an effect on prescribing. Physicians who received only meals from drug companies prescribed brand-name drugs more frequently than their counterparts who did not receive any payments from drug companies. Similarly, physicians who received payments for participating in speaking engagements prescribed brand-name drugs at a higher rate than their counterparts who received other types of payments, including honoraria, consulting fees and travel compensation, from drug companies (Grochowski Jones, & Ornstein, 2016).

9.10 The Outsourcing of Clinical Trials to Contract Research Organizations and the "Special Relationship"

The CRO entity was developed in the 1980s by entrepreneurs who realized that pharmaceutical companies were looking for research entities that were specifically engineered to absorb all of the tasks that were associated with drug testing and clinical trial research, coordination, associated deadlines and cost containment in the recruitment of increasing patient populations (Mirowski, 2011; Sismondo, 2003). The CRO has become involved in nearly every stage of the drug development and promotion processes. This observed shift in the academic medical research and pharmaceutical business structure has been interpreted in a manner that absolves the knowledge-creators of their academic research responsibilities through contemporary norms including favouring commercially confidential information (CCI) over data transparency (Lexchin, 2012; Mirowski, 2011). Where an academic physician would otherwise have analysed the data from a clinical trial, CRO employees collect the data and medical records from each clinical trial centre and analyse and interpret the results at their own company for the sponsoring company (Healy, 2012).

The success of CROs as, now, their own industry is owed to these companies' abilities to "…offer targeted drug expertise, timely clinical trial completion, and eventually 'end-to-end outsourcing support for all phases of clinical research' at a comparatively lower cost" (Mirowski, 2011). CRO employees have the ability to alter the research protocols in a given clinical trial, thereby, augmenting the sponsoring company's bottom line (Mirowski, 2011). Furthermore, CROs have developed expertise in foreign relations and specialization in regulatory regimes in a globalized world, which allows them to coordinate clinical research internationally as required by the sponsoring company (Mirowski, 2011). These service offerings have proven useful to drug companies that opt to outsource key elements of their supply chain. Drug companies continue to outsource much of their routine drug development and clinical trials, and, therefore, externalize their financial risk and risk of obtaining unfavourable results to the CRO industry (Mirowski, 2011). CRO employees also oversee clinical trials that are performed in a multi-centre format, distributed across several clinicians who are separated by geography, potentially making uniform data collection more challenging and fragmented. However, regardless of the clinical trial format, the privatized and commodified nature of pharmaceutical research ensures that any research output that is published or otherwise disseminated is only based on that which the sponsoring company agrees to release (Mirowski, 2011).

Contract research organizations have imposed a new set of normative behaviours that are suited to controlling the R&D and marketing cycles of new medicines, rather than providing scientists and the medical community with the opportunity to construct research protocols for their research. CROs have constructed normative research practices that are suited to the cadence of corporate-privatized science in the global knowledge economy. Some research practices that have become normative in CROs include maintaining data secrecy and confidentiality, avoiding extended treatment regimens that are inconvenient and costly and are irrelevant to the research program, minimizing patient interaction and amending research protocols to achieve a favourable bottom line (Mirowski, 2011). These practices can be, undoubtedly, detrimental to the health of the patients, who are ultimately treated based on the results of research conducted in this manner.

Although clinical trial recruitment and ethical treatment of participants are not within the scope of this dissertation, it is important to mention that the ability of CROs to recruit patients internationally has cultivated continued debate about the ethical treatment of research participants. While trials conducted and managed by CROs in both developed and developing countries may foster increased access to medicines, the recruitment of subjects from third-world countries into clinical trials parallels vulnerability and poverty status. The nefarious and unethical treatment of North American clinical trial participants recruited to participate in trials conducted and managed by CROs have also been documented (DrugWatch, 2015; Elliott, 2010, 2011, 2012, 2014). The terms "guinea pigs" and "foreign bodies for sale" have been used to refer to the manner in which the recruitment of international clinical trial participants by CROs is considered within neoliberal science (Elliott, 2010; Mirowski, 2011).

There is limited access to trial data for academic scientific researchers, who are often physicians, are recruited to sign on to the publication of the research to afford its scientific credibility. The credibility of the physicians who are named as authors on these research publications disguises the fact that the authors were unlikely to have had full access to all of the data at each, or any, stage of the clinical and analytic processes (Moffatt & Elliott, 2007; Ross, Hill, Egilman, & Krumholz, 2008; Sismondo, 2009). A selling point of the CRO is that it accommodates the private industry's goals by only releasing data if and when the sponsors sanction its release (Mirowski, 2011). If a sponsor does not sanction the release of its data, then the CRO will keep it confidential through a series of restraint clauses, confidentiality provisions, publication embargoes and other legal methods to control proprietary information (Mirowski, 2011). Pharmaceutical companies have willingly utilized their legal powers to prevent or restrict disclosure. The CRO assists in these methods of restricting disclosure of almost all aspects of the clinical trial process in which the CRO plays a role (Mirowski, 2011).

Mirowski (2011) explains that some commentators have put forward the counterargument, that when clinical trials are distributed across several clinicians at various geographically located sites, it is highly unlikely that the many small decisions made by clinicians and physicians could favourably bias the results for the sponsor; however, while these sorts of decisions have always affected clinical trials, the privatization of science has insulated these decisions from both internal and external critique. Overall, the result is that the published and otherwise disseminated science is only that which has been agreed to be released by the funder (Bekelman et al., 2003; Friedberg, Saffran, Stinson, Nelson, & Bennett, 1999; J Lexchin et al., 2003; Lundh, Lexchin, Mintzes, Schroll, & Bero, 2017; Mirowski, 2011; Rochon et al., 1994; Sismondo, 2008; Stelfox, Chua, O'Rourke, & Detsky, 1998). CROs also sidestep any issues pertaining to academic freedom because they only answer to the drug companies that hire them and hold no responsibility or accountability for the accuracy of the research results (Mirowski, 2011). For example, CRO employees are not liable for product negligence for a number of reasons including that they are anonymous in the process, there is no single person or small number of people who stand firmly behind the research that they produce, and because the rate of turnover at CROs is high, employees seldom see a project through to its completion (Mirowski, & Van Horne, 2005).

This responsibility to only the drug company with which CROs have contracts further insulates the work output at CROs, which is of considerable concern, especially since CROs have gradually become responsible for clinical drug testing (Mirowski, 2011). Furthermore, it is in their contracts with drug companies that CROs will not patent any of the research tools that arise from their contracted research that has been conducted by CRO employees. Because they are, by nature, uncurious, it is the job of CROs to only provide their sponsors with predefined data and they neither receive credit for, nor contribute to, the research that comprises the body of knowledge. If employees in CROs do not comply with their roles, their employment is terminated (Mirowski, 2011). With this transformation from open science to private science, i.e., from academic centres to CROs, universities and academic research centres have become unable to maintain their mandate of science for the public good (Mirowski, 2011) to the detriment of not only the universities as publicly oriented academic research institutions, but also the scientific research that is published in medical journals and otherwise disseminated.

9.11 Informing Physicians: Health Canada as a Learned Intermediary?

Health Canada is the Canadian drug regulatory arm of the Canadian government. In order to begin the drug approval process in Canada, a sponsoring drug company files a New Drug Submission (NDS) with Health Canada. This NDS includes preclinical and clinical trial information regarding a product's safety, efficacy and quality information. Health Canada relies on the sponsoring drug company to submit its own data and interpretations of that data for review. A 2016 study by Joel Lexchin assessed the characteristics of studies that Health Canada used between 1 January 1998 and 30 June 2017 to make decisions to grant full marketing authorization, or a Notice of Compliance (NOC), to products that had been given a conditional approval, or a Notice of Compliance with Conditions (NOC/c) (Lexchin, 2018).

In order for the status of products to be upgraded from a NOC/c to NOC, the sponsoring company is required to conduct additional post market trials. If the post market trials fail to provide sufficient evidence of clinical benefit, the NOC/c could be revoked altogether, resulting in the product being removed from the market. This study found that although Health Canada tended to require a more rigorous methodology than the United States Food and Drug Administration and the European Medicines Agency, the post market studies provided to Health Canada were observational in nature, looking at rare side effects, rather than establishing efficacy of the products (Lexchin, 2018). Other limitations of the industry data on which Health Canada relied included a lack of blinding or lack of statement of blinding in almost 66 percent of studies, use of surrogate outcome measures, and a lack of gender diversity within the studies, leading to questions about how various products will affect men as compared with women (Lexchin, 2018).

Health Canada provides the Summary Basis of Decision (SBD) documents, which provide the clinical trial information that it considers when approving a new drug. A 2016 study by Roojin Habibi and Joel Lexchin found that there are significant omissions in the clinical trial information, which prevents physicians from making informed prescribing decisions. This study found that clinical trial information is presented haphazardly, without comprehensible order. Although basic clinical trial characteristics were frequently described in the SBD documents, the majority of the SBDs failed to provide information about patient trial characteristics and the benefits and risks of tested treatments. A lack of information from Health Canada leaves physicians with an information deficit regarding prescribing decisions; however, Health Canada may be left in a similar information deficit, since drug companies remain in a position to provide Health Canada with data due to the current pharmaceutical research and regulatory environment.

For instance, the RIAT initiative has shown that the participant-level clinical trial data in Study 329 on Paxil was skewed in its coding of serious adverse events, affecting the information on which Health Canada and other regulators based their approval decisions (Le Noury et al., 2015b). In fact, once reanalysed, the Study 329 data was found to have been misrepresentative of the safety and efficacy of both paroxetine. The reanalysis of the data found that there were clinically significant increases in harms associated with paroxetine, including suicidal ideation, suicidal behaviour and other aforementioned serious adverse events (Le Noury et al., 2015b). According to exhibits made public from a major lawsuit regarding Paxil called *Dolin v GlaxoSmithKline* in the United States, "[a]n accurate reading of GSK's original safety data submission to the FDA, excluding placebo run-in events, comparing the on-Paxil and on-placebo event rates to each other, **patients taking Paxil were at an 8.9 times greater risk of experiencing a suicide event than those on placebo**" (emphasis in source) (Baum Hedlund Aristei Goldman PC, 2009). On reanalysis, both drugs were found to have an increased risk of harm, while neither paroxetine nor the comparator, imipramine, showed efficacy for major depression in adolescents. The authors of the reanalysis conclude that without access to and interrogation of primary clinical trial data, published conclusions about the clinical safety and efficacy should not be read as authoritative (Le Noury et al., 2015b).

9.12 Agency Theory and Vicarious Liability in Tort: Can a Special Relationship Exist Between Pharmaceutical Companies and Consumers?

Pharmaceutical companies have a relationship with the entities to which they contract their research and writing services, but whether pharmaceutical companies have a special relationship with consumers is another matter. As mentioned, the doctor–patient (*Norberg*, 1992), parent–child (*M(K)*, 1992), First Nations groups-Crown (*Guerin*, 1984), solicitor–client (*Kitchen v Royal Air Force Association*, 1958) and bailor–bailee (*Beaman v ARTS Ltd.*, 1949) relationships have all been deemed sufficient to pass the special relationship threshold. To determine whether a special relationship can exist between pharmaceutical companies and consumers, agency theory may be applied. Given the direct link between the sponsoring company's research, its reported adverse events data, and the potential for these side effects to occur in the consumer population, there appears to be some kind of relationship between drug companies and their consumers. The question, however, remains as to whether this relationship can be recognized as a special relationship. As we see from the paroxetine case study, it is only the sponsoring pharmaceutical company that had any knowledge of the true adverse events data. Furthermore, it is only the sponsoring company that had access to this data, to the extent that it, alone, held the responsibility of disclosing its data to health regulators and people who wrote its journal manuscripts, among others.

Agency theory analyses operate to develop an understanding of the nature of the economic relationship between the "agent", a person who possesses significant control over the assets of the "principal", a person or entity who possesses the residual economic interest in those assets (Nicholls, 2005). In the GSK-paroxetine case study, the agent–principal actors would differ based on entity and role in the complex pharmaceutical supply chain. For instance, for economic purposes, GSK's marketing managers, officers and directors may function as "agents", while the shareholders function as "principals" (American Law Institute, 2006; Nicholls, 2005). The managers, officers and directors, though, still have an interest in demonstrating their competency in the form of increasing the value of the company through sales, thereby, increasing the value of the shareholders' investments. GSK's employees, including its researchers, also function as agents to GSK, as corporate principal. Further, as contracted entities, the CROs and MCCs may be viewed as "agents" for GSK as "principal". As agents, the CROs and MCCs would benefit from demonstrating their abilities to ensure their deliverables, for example, as STI quoted in its proposal for publications services to GSK. As an economic approach, agency theory proposes that managers, officers and directors who destroy value and cause a decline in share value will put their current and future jobs in jeopardy (Nicholls, 2005). Furthermore, the interests of the corporation are not to be confused with the interests of stakeholders (Peoples Department Stores Inc. (Trustee of) v Wise, 2004 [Peoples]); however, these interests may become blurred as shareholders determine their financial interest in a company. Accordingly, the acts of GSK employees, managers, officers and directors who engage in and condone clinical trial data manipulation practices appear to make practical sense in terms of their job security and duty to GSK and its shareholders.

Agents of a corporation are burdened by corporate law as stated in the *Canada Business Corporations Act* (*CBCA*) with two overarching duties: (1) a duty of care, diligence and skill, and (2) a duty of loyalty and good faith, which is described as a fiduciary duty (CBCA, s. 122(1); *Peoples* 2004). The duty of care, diligence and skill required by corporate agents involves taking steps to prevent corporate wrongdoing (Nicholls, 2005). By extension, this duty may signal to consumers that the corporation, via its agents, is taking care of protecting the safety of the consumers. This signalling is further supported by consumer safety legislation that requires that companies' products, especially in the sector of pharmaceuticals as highly regulated products, do not cause harm that can otherwise be prevented, or about which can be warned. This duty operates alongside the fiduciary duty that corporate agents possess to act in the best interests of the corporation. There may be an exception to this unwavering fiduciary duty, though, per the Declaration of Helsinki (World Medical Association, 2013), which outlines ethical principles for research involving human subjects. The general principles of the Declaration of Helsinki are addressed primarily to physicians, though, stating the responsibility that physicians have in research involving humans is first to the patients, acting in the patient's best interests and their rights when providing medical care.

Perhaps the Declaration of Helsinki is outdated or naïve in respect of the personnel who are in reality charged with conducting clinical trials, making the data coding decisions, analysis and interpretation of data. Nowhere does the Declaration of Helsinki refer to either the drug companies that conduct and analyse their own trials, or the contract research and communications organizations, which possess incredible involvement in every single step and analysis carried out on the sponsor's pharmaceutical product. Complete exclusion of consideration of these inherent and clear conflict of interest relationships ultimately places both the clinical trial participants and ultimate medication consumers in vulnerable situations in which they are beholden to the conduct of the companies throughout their global supply chains. While physicians may be included within the personnel employed by these companies, they would similarly function as "agents" to the corporate principals, presumably placing them in a closer relationship with their employers than their patients.

Under corporate liability for tort principles, the doctrine of vicarious liability dictates that a corporate principal, such as GSK, may be found liable for the negligent acts of its agents (Nicholls, 2005), which, arguably, would be its own employees, managers, officers and directors, as well as those at the CROs or MCCs, with which GSK may contract. *Bazley v Curry* [*Bazley*] (1999) relies on the *respondeat superior* doctrine, which states that a principal is responsible for the acts of its agents, to establish vicarious liability of the principal. The decision begins with the statement that "[i]t is tragic but true that people working with the vulnerable sometimes abuse their positions and commit wrongs against the very people they are engaged to help" (Bazley v Curry, 1999). Although *Bazley* is a sexual abuse case, the sentiment regarding the vulnerability of the subjects resonates similarly with the wrongful act of data fraud having been committed, causing additional risk for harms that may have otherwise been avoided. A combination of corporate liability and tort principles may, therefore, hold GSK vicariously liable for the intentional acts of its employees as agents in choosing to misrepresent paroxetine's serious adverse events profile, as well as the STI employees who were paid by GSK to fulfill certain promotional objectives in its publication management and ghostwriting of Study 329.

The relevant issue in *Bazley* concerned whether principals could be held liable for their agents' sexual assaults on the persons within their care. Similarly, the issue in the GSK-paroxetine case study is whether the principal can be held liable for the agents' intentional acts of fraudulent concealment of data for the ultimate consumers who, presumably, believe to be under the care of the company by virtue of ingesting its product, hoping for treatment. Of course, the physician is considered a learned intermediary between the company and the consumer; however, because the drug company retains control over its data and its release, physicians are not truly learned intermediaries because they do not have access to all of the relevant data (Tyler & Cooper, 1997).

The cases that confirm vicarious liability on the basis that the employee was acting to drive forward the goals of the employer assume that the employee had "ostensible" or "implied" authority to do the unauthorized act; however, this argument may be insufficient for intentional torts, unless the employer's creation or enhancement of risk can be established (*Bazley*, 1999). Whether an employer's creation or enhancement of the risk can be established depends on factors including, but not limited to whether the principal afforded the agent an opportunity to abuse his or her power, the extent to which the wrongful act may have furthered the principal's aims, the extent of power conferred on the agent in relation to the clinical trial participant, in this case, or the ultimate consumer, and the vulnerability of potential participants and consumers to be affected by the wrongful exercise of the agents' power (*Bazley*, 1999). When applying these considerations to wrongdoing by agents, it must be possible that "the employer *significantly* increased the risk of harm by putting the employee in his or her position and requiring him or her to perform the assigned tasks (*Bazley*, 1999)." Incidental considerations of time and place do not provide a strong enough connection (*Bazley*, 1999).

To apply the *Bazley* criteria, one needs to consider the business structure common to large pharmaceutical companies, such as that of GlaxoSmithKline. The business structure can be seen to create the necessary opportunity for the abuse of agents' power in the research and data dissemination processes required to sell pharmaceuticals. After all, the practice of data suppression, which can manifest as sponsorship bias, is quite common throughout the industry (Angell, 2008; Lundh et al., 2017; Smith, 2005; Spielmans & Parry, 2010). Moreover, the suppression of unfavourable clinical trial data via re-coding furthers the corporate principal's aims by promoting sales of the product with favourable data. In the GSK-paroxetine case study, the trial results indicated that the drug was no more effective than placebo (Kondro, 2004). Despite these results, internal GSK documents show GSK employees writing that the "[p]ositive data from Study 329 will be published..." and that "[i]t would be commercially unacceptable to include a statement that efficacy had not been demonstrated, as this would undermine the profile of paroxetine", showing deliberate and successful efforts by GSK employees to skew the data (Kondro, 2004).

Both clinical trial participants and the ultimate consumers of the pharmaceutical product in question are vulnerable, perhaps in differing manners, but nonetheless share the vulnerability that they both hope for the treatment to be effective, their health outcomes rely on the data collection and interpretation of said data by the sponsor, and neither they nor any other independent researcher can scrutinize the data to ensure that the recorded outcomes reflect objective reality of serious adverse events outcomes. There exists a clear power imbalance between the corporate agents and the clinical trial participants and ultimate consumers, coupled with the pressure to maximize profits during the drug's patent life, while reducing risk or cost to the company.

Norberg (1992) states that the essence of a fiduciary relationship is that one party exercises power on behalf of another, the party holding the most power on behalf of another vows to act in the best interest of the more vulnerable party. Upon a breach of the fiduciary relationship, the wronged party is favoured in remedy. A fiduciary relationship is supported by certain characteristics, which include that the fiduciary can exercise discretion or power, the fiduciary can unilaterally exercise that power or discretion to affect the legal or practical interests of the "beneficiary", and the beneficiary is particularly vulnerable or at the mercy of the fiduciary, who holds the power (1992). *Norberg* describes these characteristics in the context of the physician–patient relationship; however, these traits may similarly be established in the GSK-paroxetine case study.

Based in the United Kingdom and headquartered in Brentford, London, GlaxoSmithKline Finance PLC (public limited company), agreed to pay the fines for its American subsidiary, SmithKline Beecham, at the time of the fraud, which continued until the 2000 Glaxo Wellcome and SmithKline Beecham merger, forming GlaxoSmithKline. Although the parent company paid the fraud settlement, the depths of the fraudulent actions continue to be realized today, for example, with the re-publication of Study 329 in 2013. In this vein, the 2013 ONSC *Choc v Hudbay Minerals Inc.* [*Choc*] (2013) motion decision was precedent-setting in that it imposed liability on the parent company, Hudbay Minerals Inc., for the alleged violations in tort that its Guatemalan subsidiaries' employees caused to the indigenous population, as well as the death of a community leader allegedly caused by the actions of the company, in the particular mining region in Guatemala (*Choc*, 2013). In the motion to strike, although the parent company argued that it had no legal responsibility for the actions of its Guatemalan subsidiary, the court determined that the corporate veil could be pierced and the parent company was responsible, in part, for the actions of the subsidiary.

Although liability has not yet been established on the facts in *Choc*, similar to the way that Hudbay Minerals Inc. could be held liable for the negligent supervision of the way its subsidiaries directed security forces in Guatemala, in the drug company-CRO research context, the parent or sponsoring company could in theory be liable for negligent supervision of the way its subsidiaries, employees, or contractors directed its research and data interpretations to be conducted. Even if CROs, for example, were to intentionally falsify results, a plaintiff would not have to show that the sponsoring company was actually aware of the specific falsification as long as the plaintiff could show that the sponsoring company was generally aware that this could happen and took no steps to prevent it or ensure that it did not happen. In this situation, the company would not be able to rely on wilful blindness.

If the fiduciary is considered to be the corporate agents, or researchers, who oversaw, re-coded, analysed, and interpreted the data of Study 329, they will have exercised their discretion or power in making these decisions accordingly. The corporate agents, as fiduciaries to the study participants and ultimate consumers of paroxetine, were able to unilaterally exercise their discretion and power to affect the legal and practical interests of the clinical trial participants and ultimate consumers. For example, suppression of the suicidality data affected the practical and legal interests of the ultimate consumers. The practical interests of the ultimate consumers of paroxetine would have been negatively affected because if they experienced suicidality while taking paroxetine, they would not have known it to be a side effect and may inquire about receiving a higher prescription of the selective serotonin reuptake inhibitor (SSRI), rather than, perhaps, adjusting their dosage with their physician or changing medications altogether. Additionally, the legal interests of these consumers will have been negatively affected because the suppression of suicidality data may have prevented or delayed discoverability of the potential for harm. If an ultimate consumer taking paroxetine experienced suicidal ideations or suicidality and sued the company beyond the expiration of the limitation period, the corporate principal may argue discoverability; however, the fraudulent concealment of the relevant data will have prevented the consumer from exercising his or her legal rights. Finally, both the Study 329 clinical trial participants and ultimate consumers are inherently in a vulnerable position, the former ingesting a medication that has not yet been fully studied and the latter ingesting a medication for which not all of the relevant data had been released.

A reasonable person would consider intentional acts of the misrepresentation of data from clinical trials to regulators, medical professionals and the public to be unconscionable. The unique circumstances of privatized pharmaceutical research and information dissemination, in the context of disclosure and accurate representation of serious adverse events outcomes in clinical trials, perhaps, provides support for the argument that drug companies are engaged in a special relationship with their ultimate consumers. The nature of the umbrella of power, possessed by drug companies, over their supply chain and, particularly, over their own research and publication elements of the pharmaceutical supply chain, supports the direct link and, therefore, a special relationship, between drug companies and the ultimate consumers of their products.

9.13 Conclusion

Establishing the existence of a special relationship between drug companies and their ultimate consumers is an interdisciplinary exercise, which appreciates the intersectionality between various areas of law and scholarship. The pharmaceutical business structure, while vast and global in nature, may not be able to preclude drug companies from engaging in "special relationships" with their consumers, especially when no intermediary inspection of their data, its analysis and results can be conducted. The link between drug companies and their consumers arguably remains as long as the drug is on the market, consumers continue to consume the medication, and there is a potential for known side effects to occur. Fraudulent concealment, as an equitable doctrine, aims to grant relief in equity in accordance with both statute and common law. This analysis provides an avenue by which the first element of the fraudulent concealment test, that is, establishing the existence of the special relationship, can be considered in the context of products liability and business law in pharmaceuticals cases, so that the second and third elements of the test can be subsequently determined. This analysis is unique in its application to the evidentiary record of GSK's business practices, but is transferrable to other products liability cases wherein the learned intermediary doctrine is of little assistance, intentional data misrepresentation has been established, and a special relationship must be established between the corporate principal and ultimate consumer for litigation to continue in equity.

When data is withheld, misrepresented, or mischaracterized, and a consumer can make a products liability claim in negligence or intentional tort, the special relationship is a factor in the consideration of responsibility, duty and relationship between the company and consumer. It is both reasonable and foreseeable to a company that its consumer would rely on its representation of its products and risks associated with its products. Similarly, a consumer would rely on a company's representations of its products is both reasonable and foreseeable.

By the estimation of this chapter, the special relationship may reasonably be cautiously applied to the facts to the relationship between drug companies and their consumers based on the foregoing analysis. In the relationship between drug companies and their consumers, there is a clear comparative differential in power, where the company holds both financial and informational powers, to which the consumer is not privy. In cases where material facts such as known side effects are withheld, misrepresented, or mischaracterized by the sponsoring company, the learned intermediary doctrine cannot apply; for in order for the doctrine to apply, the intermediary must have access to the information that is necessary to make an informed decision, or assist the consumer in making an informed decision. When neither the regulator, nor physicians, have access to data that is withheld, misrepresented, or mischaracterized, by definition, they cannot be learned intermediaries. Furthermore, based on *Hollis*, the nature and scope of the manufacturer's duty increases with the level of danger associated with the ordinary use of the product. Therefore, if serious adverse events such as suicidality, violent suicidality, or violence against the person or others are associated with the ordinary use of a drug, this would trigger the need for sufficient detail about these risks to be released by the sponsoring company.

The potential ineffectiveness of intermediaries without complete data on the facts exacerbates the vulnerability of consumers who take the affected drug products. Although consumers place trust and confidence in their prescribing physicians, the physicians place their trust and confidence in the information sources including the regulators, clinical practice guidelines, medical education and sponsoring companies for prescribing information. When important adverse events data is withheld, misrepresented, or mischaracterized, consumers arguably become more vulnerable to the risks associated with the affected products, shrinking the relational distance between the consumer and sponsoring company. Courts would unlikely make the leap of deciding that there exists a fiduciary relationship between companies and consumers considering the statement in *Imperial Tobacco*, that the element of vulnerability elevates the relationship to fiduciary duty; however, it is possible for the courts to decide the development of a special relationship where the facts meet the available criteria for such a relationship. Establishing such a special relationship would likely require more study of the special relationship and its application. Dr. Marcia Angell, former editor of the *New England Journal of Medicine* has said that drug companies can actually "change the way medicine is practiced" by failing to disclose unfavourable research results. By extension, in pharmaceutical products liability cases, by failing to disclose negative research results, drug companies can similarly change the cases are brought and the decisions of the courts in products liability and pharmaceutical harm cases.

Acknowledgements I would like to thank Dr. Philip Girard, professor and legal scholar at Osgoode Hall Law School in Toronto, Canada, for his helpful comments on a draft of this chapter.

Appendix 9.1. Proposal to SmithKline Beecham for a Journal Article by Scientific Therapeutics Information, Inc.[1]

FILE COPY

ADOLESCENT DEPRESSION

STUDY 329

Proposal for a Journal Article

Proposed to:

Ivan Gergel, MD
SmithKline Beecham Pharmaceuticals

ok direct bill

Proposed by:

Sally K. Laden, MS
and
John A. Romankiewicz, PharmD
Scientific Therapeutics Information, Inc

April 3, 1998

57881040398

PAR001373269

PAR0013732

[1]Laden, SK; Romankiewicz, JA. Adolescent Depression Study 329: Proposal for a Journal Article. 1998 April 03. POGO Investigation on Ghostwriting Collection. Unknown. https://www.industrydocuments.ucsf.edu/docs/npfw0217.

TABLE OF CONTENTS

Page

Purpose 1

Description 2

Target Journal 3

Authors 4

Services 5

Time and Events 7

Costs 8

Payment Schedule, Authorization 9

1

PURPOSE

The purpose of this proposal is to outline the timing, services, and costs associated with writing a manuscript for journal publication that will review the results of a study on "PAXIL in the Treatment of Adolescent Depression," Study 329. Data from Study 329, published literature, and unpublished data to be provided by SmithKline Beecham Pharmaceuticals form the basis of this article.

The article will be developed for a psychiatry journal. A target journal is recommended, however, the decision on which journal to submit the paper to will be determined after review of the study report and after consultation with the primary author and the sponsor.

57881040398

PAR001373271

PAR001373271

2

DESCRIPTION

The manuscript that is developed will be between 20 and 24 double-spaced typewritten pages and will include figures, tables, and references as appropriate. The manuscript will be designed to conform to the style of the journal selected.

A complete outline of the article will be provided upon review of the data with which we will be working.

3

TARGET JOURNAL

There are several potential journals for an article of this nature. The target journal for publication of this manuscript is the *Journal of the American Medical Association*.

57881040398

PAR001373273

PAR001373273

4

AUTHORS

The primary author of this paper will be Marty Keller, MD. Costs take into consideration that up to three (3) authors will be reviewers on the paper. Additional individuals with associated review and comments will be assessed additional charges.

STI will work with Jim McCafferty of SmithKline Beecham as our primary contact for acquisition of data and coordination of reviews by SmithKline Beecham.

5

SERVICES

STI editors will write up to six drafts of the manuscripts with the sixth being the journal submission draft. STI will also work with the sponsor and authors to ensure that the most appropriate journal for publication of the manuscript has been selected.

We will need to discuss journal submission, coordination of the review process, and number of draft(s) with the authors at the start of the review process to minimize untimely delays in the review and commentary process.

STI will develop up to six drafts: Draft I is the initial draft that will be reviewed by the sponsor. Comments on Draft I will be incorporated into Draft II, which will be sent to the primary author and the sponsor for comments. Draft III will incorporate any comments and will be sent to the same reviewers plus up to two (2) additional non-SB authors for review and critique. Draft IV will be sent to all authors and sponsor for comment. Draft V is the pre-journal submission draft and will be sent to all authors and sponsor for final review and approval prior to developing the journal submission package. Draft VI is the journal submission draft prepared for the journal, for which art work will be professionally drawn and the manuscript styled according to the selected journal. Proper referencing and obtaining permissions for use of previously-published data are included. A submission package will be prepared that will be forwarded to the primary author to send to the journal.

57881040398

PAR001373275

PAR0013731

6

SERVICES (continued)

STI will provide all necessary resources to complete this manuscript including writing, editing, library research and retrieval, copy editing, proofreading, word-processing, art work, and the needed coordination with author(s), sponsor, and journal. Also includes distribution of the final draft to the listed authors as a courtesy.

7

TIME AND EVENTS

The manuscript will be completed according to the time and events schedule outlined below. The time from submission to the journal to publication in the journal may take 4 to 8 months.

TIME	EVENT
2nd quarter	Approval to proceed provided
2 months after receipt of study report	Draft I to SB*
3 weeks after receipt of comments	Draft II to primary author and SB
3 weeks after receipt of comments	Draft III to authors and SB
3 weeks after receipt of comments	Draft IV to authors and SB
3 weeks after receipt of comments	Draft V to authors and SB for final approval and signoff for release to journal
3 weeks after receipt of signoff	Draft VI, journal submission package sent to primary author to forward to journal
4-8 months after journal submission	Paper appears in print (estimate)

*We assume that these author and sponsor review periods are limited to no longer than 3 weeks per review.

57881040398

PAR001373277

COSTS

Editorial Development:	Includes all services as outlined under "Services," including reviews and coordination with up to three non-SB reviewers plus one SB reviewer.	COST:	$16,750
Honorarium:	Not needed.		
Mailings:	A pass-through cost for overnight mailings and author reminders to return their comments is estimated.	COST (Estimated):	$500 *
		TOTAL:	$17,250

Costs exclude: More than 6 drafts; greater than 4 reviewers (SB and non-SB combined), international correspondence; page costs if charged by the publishing journal; costs of journal reprints; copyright royalties, if assessed, in the event the article uses any copyrighted figures or tables belonging to other publishers; working period beyond March 1, 1999; travel expenses for STI editor if needed to visit the sponsor or author (charged at cost plus daily fee for editor time). If any of these are incurred, cost estimates will be provided at the time of the need.

* Pass-through cost.

PAR001373278

9

PAYMENT SCHEDULE, AUTHORIZATION

Upon initiation of work	$8,500
Upon submission of Draft III	$5,125
Upon submission to the journal	$3,625
TOTAL:	$17,250

Authorization to proceed for
SmithKline Beecham Pharmaceuticals:

Signature

Print name, title

Date

57881040398

PAR001373279

PAR001373279

SB SmithKline Beecham **Audit request**

Important financial documents attached.
Return immediately to Accounts Payable, FP0825.

To	Ivan Gergel	Extension		Mailcode	4P2410
From	Trudy D.	Extension	4346	Date	10/2

Company policy requires more information to process the invoice(s) listed below. Please respond to the box(es) checked below. Vendor cannot be paid unless this form and the original invoice are returned to Accounts Payable, FP0825.

Vendor	Invoice	Purchase Order #	Amount
Scientific Therapeutics	#3079		$8500.00

RUSH!

- [ ] Please resubmit on a check request.
- [ ] Please approve the invoiced price.
- [ ] Does this invoice complete the PO? ☐ yes ☐ no
- [x] Has work been completed as invoiced? ☐ yes ☐ no
- [ ] Has material been received as listed on the invoice? ☐ yes ☐ no
- [ ] Please inform your receiving department that material has been received.
- [ ] This material or service requires a purchase/change order through purchasing department. Please supply.
- [x] Complete departmental cost center/object code required. Please supply subsidiary/subledger if needed.
- [ ] Please obtain proper approval, as listed in the chart of accounts.
- [ ] Higher approval required. Your approval limit is ________
- [ ] Please attach copy of ________
- [ ] Please submit original invoice/statement, photocopy not acceptable.
- [ ] Invoice differs from PO. Change order needed.
- [ ] Invalid PO number. Please advise.
- [ ] RUSH – Return immediately! CASH DISCOUNT INVOLVED/PAST DUE

Full signature required. Initials not accepted by auditors.

Signature [illegible] Date

Distribution White - Addressee, Return to Accounts Payable; Yellow - Accounts Payable 2A-46 (Rev.8/20/96)

PAR001373280

PAR001373280

References

Abramson, J., & Starfield, B. (2005). The effect of conflict of interest on biomedical research and clinical practice guidelines: Can we trust the evidence in evidence-based medicine? *Journal of the American Board of Family Medicine, 18*(5), 414–418.

American Law Institute. Restatement (Third) of Agency (2006).

Angell, M. (2008). Industry-sponsored clinical research: A broken system. *Journal of the American Medical Association, 300*(9), 1069–1071.

Aursnes, I., & Klemp Gjertsen, M. (2008). Common adverse events associated with an SSRI: Meta analysis of early paroxetine data. *Pharmacoepidemiology and Drug Safety, 17,* 707–713.

Aursnes, I., Tvete, I. F., Gassemyr, J., & Natvig, B. (2005). Suicide attempts in clinical trials with paroxetine randomised against placebo. *BMC Medicine, 3*(14), 5.

Bass, A. (2008). *Side Effects: A prosecutor, a whistleblower, and a bestselling antidepressant on trial.* Algonquin Books. Chapel Hill, North Carolina: Algonquin Books of Chapel Hill.

Baum Hedlund Aristei Goldman PC. (2009). GSK's knowledge of the association between Paxil and an increased risk of suicidality in adults. Retrieved from https://baumhedlundlaw.com/pdf/2-Forst-Pltfs-Add-Facts-Preemption_Redacted.pdf.

Bazley v Curry, (1999) 2 SCR 534.

Beaman v ARTS Ltd., (1949) 1 KB 550.

Bekelman, J. E., Li, Y., & Gross, C. P. (2003). Scope and impact of financial conflicts of interest in biomedical research: a systematic review. *Journal of the American Medical Association, 289*(4), 454–465.

Bell, N., Connor Gorber, S., Tonelli, M., Pottie, K., Singh, H., Joffres, M., et al. (2013). From ABCs to GRADE: Canadian task force on preventive health care's new rating system for clinical practice guidelines. *Canadian Family Physician, 59,* 1282–1289.

Bero, L., Oostvogel, F., Bacchetti, P., & Lee, K. (2007). Factors associated with findings of published trials of drug-drug comparisons: Why some statins appear more efficacious than others. *PLoS Medicine, 4*(6), 1001–1010.

BG Checo International Ltd. v British Columbia Hydro and Power Authority, (1993) 1 SCR 12.

BMJ. (2013). Restoring invisible and abandoned trials: A call for people to publish the findings. *BMJ, 346,* f4223.

Brix Bindslev, J., Schroll, J., Gotzsche, P., & Lundh, A. (2013). Underreporting of conflicts of interest in clinical practice guidelines: Cross sectional study. *BMC Medical Ethics, 14*(19), 1–7.

Campbell, E., Rao, S., DesRoches, C., Lezzoni, L., Vogeli, C., Bolcic-Jankovic, D., et al. (2010). Physician professionalism and changes in physician-industry relationships from 2004–2009. *Archives of Internal Medicine, 170*(20), 1820–1826.

Choc v Hudbay Minerals Inc., (2013) ONSC 1414.

Chopra, S. (2003). Industry funding of clinical trials: Benefit or bias? *Journal of the American Medical Association, 290*(1), 113–114.

Choudhry, N. K., Stelfox, H. T., & Detsky, A. S. (2002). Relationships between authors of clinical practice guidelines and the pharmaceutical industry. *Journal of the American Medical Association, 287*(5), 612–617.

Cosgrove, L., Bursztajn, H. J., Krimsky, S., Anaya, M., & Walker, J. (2009). Conflicts of interest and disclosure in the American psychiatric association's clinical practice guidelines. *Psychotherapy and Psychosomatics, 78,* 228–232.

Crocker v Sundance Northwest Resorts Ltd., (1988) 1 SCR 1186.

Dinnes, J., Hewison, J., Altman, D., & Deeks, J. (2012). The basis for monitoring strategies in clinical practice guidelines: A case study of prostate-specific antigen for monitoring prostate cancer. *Canadian Medical Association Journal, 184*(2), 169–177.

Donoghue v Stevenson, (1932) AC 562.

DrugWatch. (2015). Big Pharma's role in clinical trials. Retrieved March 31, 2016, from https://www.drugwatch.com/manufacturer/clinical-trials-and-hidden-data/.

Elliott, C. (2010). *White coat black hat: Adventures on the dark side of medicine*. Boston: Beacon Press.
Elliott, C. (2011). How the FDA got the Markingson case wrong. Retrieved March 31, 2016, from http://www.thehastingscenter.org/Bioethicsforum/Post.aspx?id=5147.
Elliott, C. (2012). Dan Markingson investigation: A referenced summary of the Dan Markingson case. Retrieved May 28, 2015, from markingson.blogspot.ca.
Elliott, C. (2014). Psychiatric research abuse at the University of Minnesota. *Turkish Journal of Bioethics*, *1*(1), 38–43. Retrieved from http://www.journalagent.com/tjob/pdfs/TJOB_1_1_38_43.pdf.
Enyinda, I. C., Briggs, C., & Bachkar, K. (2009). Managing risk in pharmaceutical global supply chain outsourcing: Applying analytical hierarchy process model. *Proceedings of ASBBS*, *16*(1).
Friedberg, M., Saffran, B., Stinson, T., Nelson, W., & Bennett, C. (1999). Evaluation of conflict of interest in economic analyses of new drugs used in oncology. *Journal of the American Medical Association, 282*(15), 1453–1457.
Fugh-Berman, A., & Dodgson, S. (2008). Ethical considerations of publication planning in the pharmaceutical industry. *Open Medicine, 2*(4), 121–124.
Gagnon, M. (2009). *The nature of capital in the knowledge-based economy: The case of the global pharmaceutical industry*. York University.
Giroux Estate v Trillium Health Centre, (2005). CanLII 1488 (ON CA).
Graham, D. J. (2004). Testimony of David J. Graham, MD, MPH, November 18, 2004.
Grande, D., Frosch, D. L., Perkins, A. W., & Kahn, B. E. (2009). Effect of exposure to small pharmaceutical promotional items on treatment preferences. *JAMA Internal Medicine, 169*(9), 887–893.
Grochowski Jones, R., & Ornstein, C. (2016). Matching industry payments to Medicare prescribing patterns. *ProPublica*.
Guerin v R, (1984) 2 SCR 335.
Guyatt, G., Akl, E. A., Hirsh, J., Kearon, C., Crowther, M., Gutterman, D., ... Schunemann, H. (2010). The vexing problem of guidelines and conflict of interest: a potential solution. *Annals of Internal Medicine*, *152*(11), 738–741. Retrived from https://annals.org/aim/fullarticle/745809.
Harrison, D. (2013). *The Law of Product Warnings and Recalls in Canada*. Toronto: LexisNexis Canada.
Healy, D. (2012). *Pharmageddon*. Berkeley/Los Angeles: University of California Press.
Hollis v Dow Corning Corp., (1995) 4 SCR 634.
Hugos, M. (2011). *Essentials of supply chain management*. Hoboken: Wiley.
IFPMA and ICH. (2013). *MedDRA term selection: Points to consider: ICH-endorsed guide for MedDRA users*.
Institute of Medicine [IOM]. (2011). *Clinical practice guidelines we can trust*. In R. Graham, M. Mancher, D. Miller Wolman, S. Greenfield, & E. Steinberg (Eds.), Washington, D.C.: The National Academies Press.
Keller, M. B., Ryan, N. D., Strober, M., Klein, R. G., Kutcher, S. P., Birmaher, B., McCafferty, J. P. (2001). Efficacy of paroxetine in the treatment of adolescent major depression: A randomized, controlled trial. *Journal of the American Academy of Child & Adolescent Psychiatry*, *40*(7), 762–772.
Kitchen v Royal Air Force Association, (1958) 1 WLR 563.
Kondro, W. (2004). Drug company experts advised staff to withhold data about SSRI use in children. *PLOS Medicine, 170*(5), 783.
Kung, J., Miller, R. R., & Mackowiak, P. A. (2012). Failure of clinical practice guidelines to meet Institute of Medicine standards: Two more decades of little, if any, progress. *Archives of Internal Medicine, 172*(21), 1628–1633.
Lac Minerals Ltd. v International Corona Resources Ltd., (1989) 2 SCR 574.
Laden, S. K., & Romankiewicz, J. A. (April 3, 1998). Adolescent depression study 329: proposal for a journal article. POGO Investigation on Ghostwriting Collection. Unknown. https://www.industrydocuments.ucsf.edu/docs/npfw0217.

Lambert v Lastoplex Chemicals, (1972) SCR 569.
Langer, T., Conrad, S., Fishman, L., Gerken, M., Schwarz, S., Weikert, B., Weinbrenner, S. (2012). Conflicts of interest among authors of medical guidelines: An analysis of guidelines produced by German specialist societies. *Deutsches Ärzteblatt International, 109*(48), 836–842.
Le Noury, J., Nardo, J. M., Healy, D., Jureidini, J., Raven, M., Tufanaru, C., et al. (2015a). Appendix 3: Supplementary information on suicidal and self-injurious behaviours in study 329 [posted as supplied by author]. *The BMJ, 351,* h4320.
Le Noury, J., Nardo, J. M., Healy, D., Jureidini, J., Raven, M., Tufanaru, C., et al. (2015b). Restoring Study 329: Efficacy and harms of paroxetine and imipramine in treatment of major depression in adolescence. *The BMJ, 351,* h4320.
Lexchin, J. (2012). Those who have the gold make the evidence: how the pharmaceutical industry biases the outcomes of clinical trials of medications. *Science and Engineering Ethics, 18*(2), 247–261.
Lexchin, J. (2018). Quality of evidence considered by Health Canada in granting full market authorisation to new drugs with a conditional approval: A retrospective cohort study. *British Medical Journal Open, 8*(4), e020377.
Lexchin, J., Bero, L., Djulbegovic, B., & Clark, O. (2003). Pharmaceutical industry sponsorship and research outcome and quality. *British Medical Journal, 326,* 1167–1170.
Limitations Act, (2002), SO 2002, c. 24.
Linden, A. M. (1982). *Canadian tort law* (2nd ed.). Toronto: Butterworth & Company (Canada).
Lundh, A., Lexchin, J., Mintzes, B., Schroll, J., & Bero, L. (2017). Industry sponsorship and research outcome. *Cochrane Database of Systematic Reviews.*
Lundh, A., Sismondo, S., Lexchin, J., Busuioc, O., & Bero, L. (2012). Industry sponsorship and research outcome (review). *Cochrane Database of Systematic Reviews, 12*(MR000033).
McHenry, L. B., & Jureidini, J. N. (2008). Industry-sponsored ghostwriting in clinical trial reporting: A case study. *Accountability in Research: Policies and Quality Assurance, 15*(3), 152–167.
McInerney v MacDonald, (1992) 2 SCR 138.
Mendelson, T. B., Meltzer, M., Campbell, E. G., Caplan, A. L., & Kirkpatrick, J. N. (2011). Conflicts of interest in cardiovascular clinical practice guidelines. *Archives of Internal Medicine, 171*(6), 577–585.
Mirowski, P. (2001). Re-engineering scientific credit in the era of the globalized information economy. *First Monday, 6*(12).
Mirowski, P. (2011). *Science-mart: Privatizing American science.* Cambridge, MA: Harvard University Press.
Mirowski, P., & Van Horne, R. (2005). The contract research organization and the commercialization of scientific research. *Social Studies of Science, 35*(4), 503–548.
M(K) v M(H), (1992) 3 SCR 6.
Moffatt, B., & Elliott, C. (2007). Ghost marketing: Pharmaceutical companies and ghostwritten journal articles. *Perspectives in Biology and Medicine, 50*(1), 18–31.
Moynihan, R. (2003). Who pays for the pizza? Redefining the relationships between doctors and drug companies. 1: Entanglement. *British Medical Journal, 326*(1189–1192).
Neuman, J., Korenstein, D., Ross, J., & Keyhani, S. (2011). Prevalence of financial conflicts of interest among panel members producing clinical practice guidelines in Canada and United States: Cross sectional study. *British Medical Journal, 343,* d5621.
Nicholls, C. C. (2005). *Corporate Law*. Toronto: Edmond Montgomery Publications Limited.
Norberg v Wynrib, (1992) 2 SCR 226.
Norris, S. L., Holmer, H. K., Ogden, L. A., Burda, B. U., & Fu, R. (2013). Conflicts of interest among authors of clinical practice guidelines for glycemic control in type 2 diabetes mellitus. *PLoS ONE, 8*(10), e75284.
Orlowski, J., & Wateska, L. (1992). The effects of pharmaceutical firm enticements on physician prescribing patterns: There's no such thing as a free lunch. *Chest, 102*(1), 270–273.
Osborne, P. H. (2015). *The law of torts* (5th ed.). Toronto: Irwin Law Inc.

Papanikolaou, G. N., Baltogianni, M. S., Contopoulos-Ioannidis, D. G., Haidich, A.-B., Giannakakis, I. A., & Ioannidis, J. P. (2001). Reporting of conflicts of interest in guidelines of preventive and therapeutic interventions. *BMC Medical Research Methodology, 1*(3).

Peek v Gurney, (1873) LR 6 HL 377.

Peoples Department Stores Inc. (Trustee of) v Wise, (2004) SCC 68 (CanLII), 3 SCR 461.

Queen v Cognos Inc., (1993) 1 SCR 87.

Rochon, P. A., Gurwitz, J. H., Simms, R. W., Fortin, P. R., Felson, D. T., Minaker, K. L., et al. (1994). A study of manufacturer-supported trials of nonsteroidal antiinflammatory drugs in the treatment of arthritis. *Archives of Internal Medicine, 154*(2), 157–163.

Ross, J., Hill, K., Egilman, D., & Krumholz, H. (2008). Guest authorship and ghostwriting in publications related to rofecoxib: A case study of industry documents from rofecoxib litigation. *Journal of the American Medical Association, 299*(15), 1800–1812.

Roulston v McKenny, (2016) ONSC 2377.

Schott, G., Pachl, H., Limbach, U., Gundert-Remy, U., Lieb, K., & Ludwig, W.-D. (2010). The financing of drug trials by pharmaceutical companies and its consequences: Part 2: A qualitative, systematic review of the literature on possible influences on authorship, access to trial data, and trial registration and publication. *Deutsches Ärzteblatt International, 107*(17), 295–301.

Shnier, A., Lexchin, J., Romero, M., & Brown, K. (2016). Reporting of financial conflicts of interest in clinical practice guidelines: a case study analysis of guidelines from the Canadian Medical Association Infobase. *BMC Health Services Research, 16*, 383.

Sismondo, S. (2003). *An introduction to science and technology studies*. Malden, Massachusetts: Wiley-Blackwell.

Sismondo, S. (2007). Ghost management: how much of the medical literature is shaped behind the scenes by the pharmaceutical industry? *PLoS Medicine, 4*(9), 1429–1433.

Sismondo, S. (2008). Pharmaceutical company funding and its consequences: a qualitative systematic review. *Contemporary Clinical Trials, 29*(2), 109–113.

Sismondo, S. (2009). Ghosts in the machine. *Social Studies of Science, 39*, 109–133.

Smith, R. (2005). Medical journals are an extension of the marketing arm of pharmaceutical companies. *PLoS Medicine, 2*(5), 364–366.

Spielmans, G., & Parry, P. (2010). From evidence-based medicine to marketing-based medicine: evidence from internal industry documents. *Journal of Bioethical Inquiry, 7*(1), 13–29.

Spurling, G., Mansfield, P., Montgomery, B., Lexchin, J., Doust, J., Othman, N., et al. (2010). Information from pharmaceutical companies and the quality, quantity, and cost of physicians' prescribing: A systematic review. *PLoS Medicine, 7*(10), e1000352.

Stelfox, H. T., Chua, G., O'Rourke, K., & Detsky, A. S. (1998). Conflict of interest in the debate over calcium-channel antagonists. *New England Journal of Medicine, 338*, 101–106.

Stewart v Pettie, (1995) 1 SCR 131.

Stikeman Elliott LLP. (n.d.). Overview of Product Liability Law. In *The law of product warnings and recalls in Canada*. Retrieved from https://www.stikeman.com/-/media/files/kh-blog-resources/product-liability/lawproductwarningsrecallsch1.ashx.

Symm, B., Averitt, M., Forjuoh, S., & Preece, C. (2006). Effects of using free sample medications on the prescribing practices of family physicians. *Journal of the American Board of Family Medicine, 19*(5), 443–449.

The United States Department of Justice. (2012). GlaxoSmithKline to plead guilty and pay $3 billion to resolve fraud allegations and failure to report safety data. Retrieved from https://www.justice.gov/opa/pr/glaxosmithkline-plead-guilty-and-pay-3-billion-resolve-fraud-allegations-and-failure-report.

Tyler, B. J., & Cooper, R. A. (1997). Blinded by the hype: Shifting the burden when manufacturers engage in direct to consumer advertising of prescription drugs. *Vermont Law Review, 1073*, 1075.

World Medical Association. (2013). *WMA Declaration of Helsinki—Ethical principles for medical research involving human subjects*. Report. Retrieved from https://www.wma.net/policies-post/wma-declaration-of-helsinki-ethical-principles-for-medical-research-involving-human-subjects/%3E.

Yeh, J., Franklin, J., Avorn, J., Landon, J., & Kesselheim, A. (2016). Association of industry payments to physicians with the prescribing of brand-name statins in Massachusetts. *Journal of the American Medical Association Internal Medicine*, (online first).

Zinner, D., Bolcic-Jankovic, D., Clarridge, B., Blumenthal, D., & Campbell, E. (2009). Participation of academic scientists in relationships with industry. *Health Affairs, 28*(6), 1814–1825.

Adrienne Shnier Hons. B.A., M.A., Ph.D., J.D., is an Adjunct Professor in the School of Health Policy & Management and Global Health Programs in the Faculty of Health at York University in Toronto, Canada. Dr. Shnier completed her M.A. (2012) and Ph.D. (2016) in Health Policy & Equity in the School of Health Policy & Management. She subsequently completed her J.D. (2019) at Osgoode Hall Law School, where she was awarded the Sir Jack Jacob Prize in the Administration of Justice for the highest standing in Administration of Civil Justice: Class Actions. Dr. Shnier also served as the Editor-in-Chief of the Osgoode Health Law Association (2016–2018). She is a Research Fellow alumna of the Pharmaceutical Policy Research Collaboration (PPRC). Dr. Shnier's research focus is pharmaceutical policy, regulation, education, and promotion, health policy, governance, public policy, financial conflict of interest relationships, fraud, and law. She is author or co-author on peer-reviewed articles in health, policy, and law journals, and has presented her research, both in Canada and internationally.

Chapter 10
Developing Governance of Procurement Department in Hospitals

Kıymet Tunca Çalıyurt

If the facts don't fit the theory, change the facts.
—Albert Einstein (1879–1955).

Abstract In for-profit and nonprofit companies, procurement and accounting departments are very tricky in terms of fraud. Transactions in these two units' and their relationship should be audited meticulously and continuously by auditors. If the hospital management doesn't like to hire an internal auditor or certified fraud examiner, they should be ready to face. As in the other sectors, various irregularities can also be seen in health institutions. Purchasing and accounting departments are the leading departments in the health companies where fraud activities happen frequently. It can be observed that the employees have misconduct towards the companies they work for. Especially the employees who are not happy with their salary think she/he works more than other employees may commit the crime. There is a need for a rigorous and practical audit and control mechanism in the purchasing department where fraudulent incidents can be seen in many different ways. In this chapter, the procurement procedure in hospitals, a procedure in procurement departments, fraud prevention techniques in the procurement department, and cases from Turkey are discussed in the chapter.

Keywords Hospital · Fraud · Purchasing · Fraud · Governance

K. T. Çalıyurt (✉)
Trakya University, Edirne, Turkey
e-mail: kiymetcaliyurtconference@gmail.com

K. T. Çalıyurt (ed.), *Integrity, Transparency and Corruption in Healthcare & Research on Health, Volume I*, Accounting, Finance, Sustainability, Governance & Fraud: Theory and Application, https://doi.org/10.1007/978-981-15-1424-1_10

10.1 Introduction

"Purchasing is indeed strategic" (Carter & Narasimhan, 1996, p. 20). For years, the purchasing literature has debated the importance and contribution of purchasing to firm strategy and performance (Carter & Monczka, 1978; González-Benito, 2007; Reck & Long, 1988 cited from Schütz, Kässer, Blome, & Foerstl, 2019). Since around the early 1990s, numerous reports have indicated that in industrialized countries, healthcare managers must control cost increases much better, without sacrificing quality and service (Naylor, 1999; Ford & Scanlon, 2007; Roehrig, 2011; Hadad, Hadad, & Simon-Tuval, 2013 cited from Nollet, Beaulieu, & Fabbe-Costes, 2017) however purchasing procedures and regulations are different in industrialized countries and others. The success of purchasing depends on many indicators like

- Compliance Rate: Understand if suppliers fulfill your requirements.
- Number of Suppliers: Track your level of dependency towards your suppliers.
- Purchase Order Cycle Time: Know who to address your urgent orders to.
- Supplier Availability: Measure suppliers' capacity to respond to demand.
- Supplier Defect Rate: Evaluate your suppliers' individual quality.
- Lead Time: Understand the total time to fulfill an order.
- Cost of Purchase Order: Control the internal costs incurred by each purchase.
- Procurement Cost Reduction: Streamline the tangible cost savings.
- Procurement Cost Avoidance: Avoid potential extra costs in the future.
- Procurement Rate of Investment: Determine the profitability of investments (Datapine, 2019).

Vian, Savedoff, and Mathisen (2010) listed corrupt practices and their impact in the health sector. They also mentioned about "procurement corruption" as follows:

Explanation: Encompasses many types of abuse including bribes, kickbacks, fraudulent invoicing, collusion among suppliers, failure to audit performance on contracts, etc.
Impact: Procurement corruption raises the price paid for goods or services, thus increasing inefficiency; goods and services may not even be needed, may not be delivered, or maybe of substandard quality (Vian, Savedoff, & Mathisen, 2010).

Corruption is an ancient problem (Bardhan, 1997) and "corporate fraud examination" is a new subtitle under the management science. Unfortunately most of the managers in the health sector don't familiar what kinds of fraud they face with in the working environment.

When we talk about "fraud examination, detection and reporting", first of all, we should determine its place in the Fraud Tree to decide[1,2];

[1]Occupational frauds are those in which an employee, manager, officer, or owner of an organization commits fraud to the detriment of that organization. The three major types of occupational fraud are: Corruption, Asset Misappropriation, and Fraudulent Statements (ACFE, https://www.acfe.com/fraud-tree.aspx).

[2]When business frauds are analyzed, it is ascertained that three components come together when committing the white-collar crime. These are pressure, opportunity, and justification that constitute the "fraud triangle." (Özkul & Pamukçu, 2012).

- how to detect,
- how to report, and
- how to prevent.

Warning signs in procurement: There are many in this category, but in essence:

- Invoiced goods/services can't be accounted for,
- Records are nonexistent or don't match,
- Invoices share purchase order numbers or are identical in terms of value or service, and
- Total payments exceed the total purchase order or agreed amounts (Nextenders, 2016).

10.2 Developing Sustainability, Internal Control and IFRS for Hospital's Procurement Department

As in the other departments in hospitals, there are some tools that should be applied in order to prevent fraud in the purchasing department. These are internal control procedures, International Financial Reporting Standards, and sustainability.

10.2.1 International Financial Reporting Standards (IFRS)

IFRS Standards address this challenge by providing high quality, internationally recognized set of accounting standards that bring transparency, accountability and efficiency to financial markets around the world. Applying national accounting standards meant amounts reported in financial statements might be calculated on a different basis. Unpicking this complexity involved studying the minutiae of national accounting standards, because even a small difference in requirements could have a major impact on a company's reported financial performance and financial position—for example, a company may recognize profits under one set of national accounting standards and losses under another. IFRS Standards bring transparency by enhancing the international comparability and quality of financial information, enabling investors and other market participants to make informed economic decisions.

- IFRS Standards strengthen accountability by reducing the information gap between the providers of capital and the people to whom they have entrusted their money. Our Standards provide information that is needed to hold management to account. As a source of globally comparable information, IFRS Standards are also of vital importance to regulators around the world.
- And IFRS Standards contribute to economic efficiency by helping investors to identify opportunities and risks across the world, thus improving capital allocation. For businesses, the use of a single, trusted accounting language lowers the cost of capital and reduces international reporting costs (IFRS, 2019a).

Most of the hospitals in many countries apply IFRS that here we will talk about the use of IFRS for leasing transactions in hospitals' procurement departments. Hospitals buy or lease current or fixed assets (Investopedia, 2019b)[3] to maintain their target and existence. An asset (Investopedia, 2019a) is a resource with economic value that an individual, corporation, or country owns or controls with the expectation that it will provide a future benefit. Assets are reported on a company's balance sheet and are bought or created to increase a firm's value or benefit the firm's operations. An asset can be thought of as something that, in the future, can generate cash flow, reduce expenses, or improve sales, regardless of whether it's manufacturing equipment or a patent. Both current assets and fixed assets appear on the balance sheet, with current assets meant to be used or converted to cash in the short-term (less than 1 year) and fixed assets meant to be utilized for the longer term (greater than 1 year). Current assets include cash and cash equivalents, accounts receivable, inventory, and prepaid expenses. Fixed assets are depreciated, while current assets are not. If a hospital buys or leases assets in the private sector, the accounting department should apply national accounting standards or international financial reporting standards (IFRS) properly. Governmental hospitals apply governmental accounting standards.[4,5]

The objective of IFRS 16 is to report information that

(a) faithfully represents lease transactions, and
(b) provides a basis for users of financial statements to assess the amount, timing and uncertainty of cash flows arising from leases.

To meet that objective, a lessee should recognize assets and liabilities arising from a lease. IFRS 16 introduces a single lessee accounting model and requires a lessee

---

[3]In the healthcare industry, it is not uncommon for a hospital to contract with a medical device owner for the use of specific medical equipment for a major part of the economic life of the equipment. This type of arrangement is often priced in such a way that the consideration is based entirely on the hospital's ongoing purchase of "consumables," which allow the equipment to function as designed, and may have no minimum volume requirement. The medical device owner is willing to accept variable consideration in the arrangement because demand for the associated healthcare services suggests that a sufficient volume of consumables will be purchased by the hospital over the term of the contract to make the arrangement profitable (Deloitte, 2019).

[4]In some situations, tax-exempt debt is issued to finance construction of a facility, such as a plant or hospital, that is transferred to a user of the facility by lease. A lease may serve as collateral for the guarantee of payments equivalent to those required to service the tax-exempt debt. Payments required by the terms of the lease are essentially the same, as to both amount and timing, as those required by the tax-exempt debt. A lease modification resulting from a refunding by the lessor of tax-exempt debt (including an advance refunding) should be accounted for in the same manner (that is, in accordance with paragraphs 842-10-25-8 through 25-18) as any other lease modification. For example, if the perceived economic advantages of the refunding are passed through to the lessee in the form of reduced lease payments, the lessee should account for the modification in accordance with paragraph 842-10-25-12, while the lessor should account for the modification in accordance with the applicable guidance in paragraphs 842-10-25-15 through 25-17.

[5]Governmental authorities often use tax-exempt debt to finance the construction of a facility (e.g., a plant or a hospital) that they will lease to another entity. As noted above, the payment terms of the lease generally mirror, with respect to both amount and timing, "those required by the tax-exempt debt." Such a lease "may serve as collateral for the guarantee of payments equivalent to those required to service the tax-exempt debt."

to recognise assets and liabilities for all leases with a term of more than 12 months, unless the underlying asset is of low value. A lessee is required to recognize a right-of-use asset representing its right to use the underlying leased asset and a lease liability representing its obligation to make lease payments. (IFRS, 2019b) Applying IFRS in the health sector helps to develop standardization of corporate reporting however can't stop obstruct fraud in procurement department.

In Turkey, like the other developing countries, "public tender procedure" is being used for procurement. If hospital management would like to prevent fraud, inefficiency in the procurement department procedure, they should establish internal auditing department.

10.2.2 Internal Control Procedures

Today, the fundamental aims of business enterprises are to become current throughout the world and to expand in this competitive environment, and to increase the value and profit of shares. The realization of the aims of business enterprises depends on making the right decisions as well as an efficient and effective work. Right decisions are only made by means of reliable, true, relevant, well-timed information. The most effective element in the provision of these information in the business enterprise is the Internal Auditing Department working efficiently (Caliyurt, Agun, & Pamukcu, 2005). The function and department of internal control both in public and private institutions have gradually gained importance. Sarbanes Oxley Law (SOX), published in 2002 especially after the fraud experienced and Section 404 about internal auditing have proven that internal and external control is the most effective factors in the prevention of fraud at business enterprises.

There are a number of different definitions of the term internal control. For the purposes of Section 404, the great majority of companies and all the CPA firms use the definition in COSO (Committee of Sponsoring Organizations of the Treadway Commission)'s Internal Control—Integrated Framework. COSO's definition relates to all aspects of internal control, not just that over financial reporting (Caliyurt et al., 2005). Internal auditing is an independent, objective assurance and consulting activity designed to add value and improve an organization's operations. It helps an organization accomplish its objectives by bringing a systematic, disciplined approach to evaluate and improve the effectiveness of risk management, control, and governance processes (IIA, 2019).

The Committee of Sponsoring Organizations of the Treadway Commission (*COSO*) is a joint initiative of the five private sector organizations listed on the left and is dedicated to providing thought leadership through the development of frameworks and guidance on enterprise risk management, internal control, and fraud deterrence. We hope you will find the information on this site to be helpful and we welcome your input (Schandl & Foster, 2019) (Fig. 10.1).

The guide introduces healthcare organizations to COSO's widely used "*Internal Control–Integrated Framework*" and provides a roadmap to implementation to help

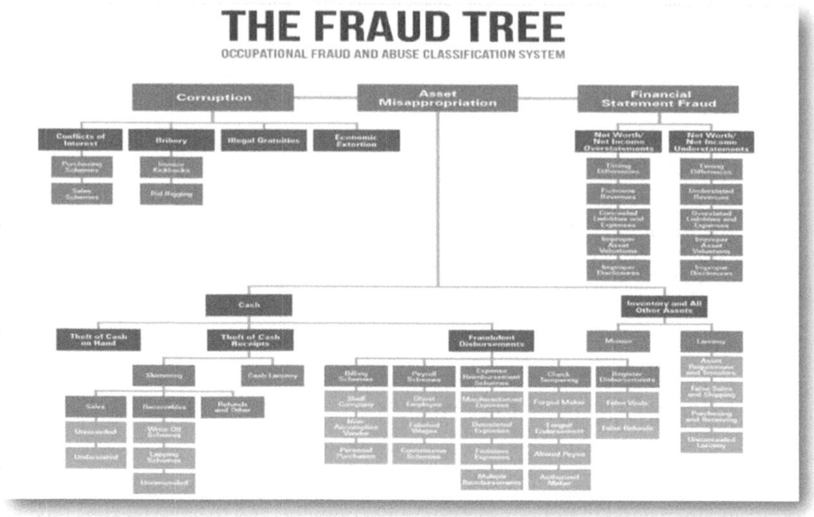

Fig. 10.1 Fraud tree

strengthen their overall governance and internal control structures. The Framework is recognized as the leading guidance for designing, implementing, and conducting internal control and assessing its effectiveness. "Effective internal control is vital to successfully weathering the ever-changing healthcare environment, and it can help mitigate many of the risks associated with the complex pressures healthcare organizations confront today," said Paul Sobel, COSO Chair. "Formally adopting the Internal Control Framework facilitates an increased understanding of the internal controls in existence and indicates where improvements should be made, resulting in reduced risk for all stakeholders." The Internal Control Framework, which can be implemented by organizations of all sizes and types, focuses on five integrated components of internal control: control environment, risk assessment, control activities, information and communication, and monitoring activities. It does this in a flexible, reliable, and cost-effective approach. "COSO provides a framework to build a fundamental foundation of internal control to ensure that organizational risks are monitored and mitigated through sound business decisions," said Bill Watts, a risk consulting partner with Crowe. "Healthcare organizations must review their environment to confirm proper controls are in place. By doing this, the organizations ensure effective and efficient operations, proper financial reporting, and compliance that supports their mission and strategy. COSO provides guidance to streamline this process." Healthcare organizations experience issues with system access and integrity, clinical documentation, coding, and billing, all of which may result in potential noncompliance with federal and state regulations—and costly mistakes Dawson and Babinchak (2019) (Fig. 10.2).

5 components	17 principles
Control environment	1. Demonstrates commitment to integrity and ethical values
	2. Exercises oversight responsibility
	3. Establishes structure, authority, and responsibility
	4. Demonstrates commitment to competence
	5. Enforces accountability.
Risk assessment	6. Specifies suitable objectives
	7. Identifies and analyzes risk
	8. Assesses fraud risk
	9. Identifies and analyzes significant change
Control activities	10. Selects and develops control activities
	11. Selects and develops general controls over technology
	12. Deploys control activities through policies and procedures
Information and communication	13. Uses relevant information
	14. Communicates internally
	15. Communicates externally
Monitoring activities	16. Conducts ongoing and/or separate evaluations
	17. Evaluates and communicates deficiencies

Source: Adapted from the COSO "Internal Control - Integrated Framework"

Fig. 10.2 Five components and 17 principles of internal control. *Source* Adapted from the COSO Internal Control–Integrated Framework (Shandal & Foster, 2019)

News from USA: What GAO[6] Found

In its December 2017 report, GAO found that the Centers for Medicare & Medicaid Services' (CMS) antifraud efforts for Medicare partially align with GAO's *2015 A Framework for Managing Fraud Risks in Federal Programs* (Framework). The Fraud Reduction and Data Analytics Act of 2015 required OMB to incorporate leading practices identified in this Framework in its guidance to agencies on addressing fraud risks (Fig. 10.3).

Why GAO Did This Study[7]

Medicare covered over 58 million people in 2017 and has a wide-ranging impact on the healthcare sector and the overall U.S. economy. However, the billions of dollars in Medicare outlays as well as program complexity make it susceptible to

[6]The U.S. Government Accountability Office.

[7]GAO has designated Medicare and Medicaid as high-risk programs, in part due to their susceptibility to improper payments—estimated to be about USD 70 billion in fiscal year 2010. Improper payments have many causes, such as submissions of duplicate claims or fraud, waste, and abuse (Ferilli, 2013).

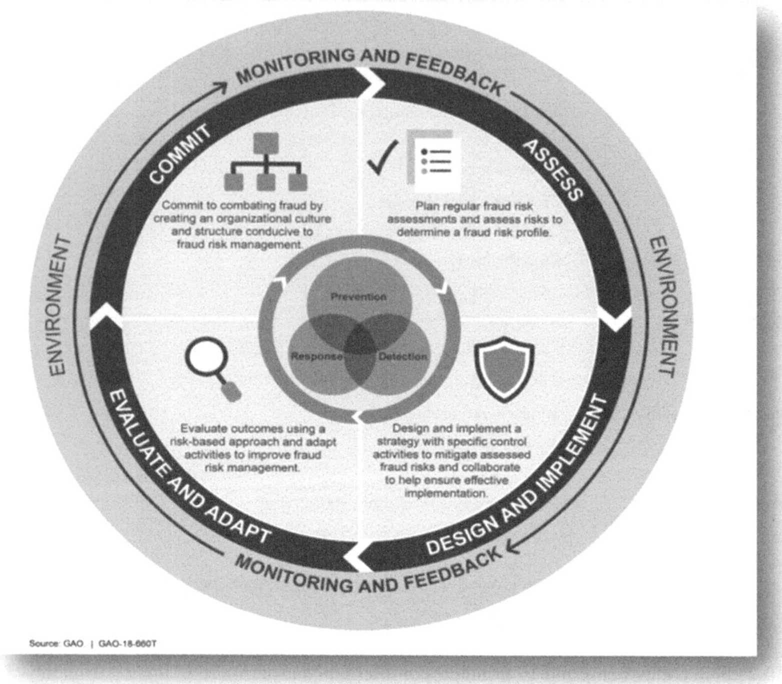

Fig. 10.3 The fraud risk management framework. *Source* GAO Highlights, 2018

improper payments, including fraud. Although there are no reliable estimates of fraud in Medicare, in the fiscal year 2017 improper payments for Medicare were estimated at about $52 billion. Further, about $1.4 billion was returned to Medicare Trust Funds in the fiscal year 2017 as a result of recoveries, fines, and asset forfeitures. In December 2017, GAO issued a report examining how CMS[8] managed its fraud risks overall and particularly the extent to which its efforts in the Medicare and Medicaid programs aligned with GAO's Framework. This testimony, based on that report, discusses the extent to which CMS's management of fraud risks in Medicare aligns with the Framework. For the report, GAO reviewed CMS policies and interviewed officials and external stakeholders.

What GAO Recommends

In its December 2017 report, GAO made three recommendations, namely that CMS;

(1) require and provide fraud-awareness training to its employees;

[8]The Centers for Medicare & Medicaid Services.

(2) conduct fraud risk assessments; and
(3) create an antifraud strategy for Medicare, including an approach for evaluation (GAO, Highlights, 2018).

Cybersecurity is another important issue for hospitals. To create a safe environment and protect data, the cybersecurity system should be bought by procurement department. Today's global business surroundings impose organizations across all industry sectors to have a secure digital infrastructure for commercial transactions. This interconnected global digital infrastructure is simply called cyberspace which includes the internet, computer systems, hardware, software and services, and digital information as a whole. It does allow e-commerce, e-government, information sharing, and trade. From the technological perspective, the connectivity, devices, and uses of today—computing tablets, home networks, smart meters, cloud computing, and social networks—have significantly made the cyberspace of today different from the past (Atkinson, Ezell, Andes, Castro, & Bennett, 2010). Cybersecurity is essentially about managing future risks and responding to current and past incidents and attacks. Managing future risk requires insight into current and future vulnerabilities and how to prevent or reduce them, the probabilities of a threat and the costs associated with potential outcomes and how to mitigate them. Responding to current and past incidents and attacks requires knowledge of what has happened, methods of preventing similar incidents from being successful in the future, and possible legal or other remedial actions against the perpetrators (Bozkus Kahyaoglu & Caliyurt, 2018). Wallace, Lin, and Cefaratti (2011) present the fact that the level of cooperation between the internal audit and information security functions is positively associated with the organization's level of compliance with the IT-related internal control requirements of the Sarbanes–Oxley Act. Similarly, Steinbart, Raschke, Gal, and Dilla (2012, 2013, 2015) propose a model of how the internal audit and information security functions could work together to support organizations to accomplish a cost-effective level of information security.

With the continuous development and improvement of the hospital system in the new era, hospital management plays an increasingly important role in the development of the hospital. Under the premise of the separation of ownership and management rights, the structure of hospital governance is formed to cope with the relationship between all stakeholders in the hospital. Internal accounting control is an important internal control measure for the development of the hospital management system. It is also a key basis to enhance the reliability and compliance of hospital financial statement. The current cost accounting system in the hospital does not meet the requirements of the new medical reform, so it is necessary to strengthen the internal accounting control system to truly strengthen the hospital cost budget. The internal accounting system of the hospital also has some shortcomings:

- First, the management awareness is not strong and the implementation of internal control management system is not enough;
- Second, the personnel related to internal control accounting lack basic knowledge and skills, and the internal control system is not perfect;

- Third, the hospital lacks the concept of financial accounting governance, resulting in a serious lack of management awareness. In view of the above problems, the hospital's internal control system needs to be improved, and the hospital's internal management policies should be improved.

Based on this, the hospital should closely link the state's financial policies and establish the governance philosophy of the relevant regulations. Besides, they should also increase the awareness of crises and improve the internal control concept of financial staff, to improve control strategies and create strict supervision of the discipline. In addition, we must also establish and improve the audit and control system. Relevant personnel in the hospital must be authorized and approved before they can handle the relevant economic activities. The scope of treatment of workers at all levels should be clarified and a mechanism of checks and balances should be gradually formed to enhance the sense of responsibility for work (Liu, 2018).

10.3 Procurement Process Outlined in Turkey and Case from the Government Hospital

According to a report published by the Ministry of Health, there are 1.514 hospitals in Turkey. 560 hospitals are private (%36) and the rest of them are governmental (Republic of Turkey, Ministry of Health, 2019). The purchasing functions of the firm are in close relationship with the control and management of this function. Therefore, in the first stage, it can be considered as a more accurate approach to explain the procurement function by explaining it fully and to analyze the administration of the procurement function.

"Procurement is the provision of the products and services needed by the enterprise at optimum cost and from reliable sources" (Scheuing, 1984: 4). Before performing any purchasing operation, it is of great importance to analyze the market or to examine the market related to the product and to determine the alternatives in the market accurately (Pooler, Pooler, & ve Farney, 2004: 89).

10.3.1 Procurement Procedure from a Government Hospital in Turkey

First step: Demand Generation

The health facility clinic management classifies and classifies the materials such as medical devices, medical consumables, and drugs, which are determined as the need during the delivery of the service, in the plans to be made during the period it serves, and presents them to the health facility management through the need assessment commissions. The needs assessment commissions decide the necessity of the requests of the health facility to the relevant clinic and make the final purchase decision.

After this stage, the clinical demand approved by the need assessment committees is brought together according to the movable codes of the requested goods and made into final demand by the relevant movable registration control authorities and opened for signature.

Second step: Technical Specification Formation Process
The clinical management that creates the demand must submit the demands to the health facility and at the same time prepare the technical specifications for the related goods in order to determine what kind of material/device/product to buy.

The technical specification shall be formally prepared in accordance with a content containing the standard form and letterhead of the health facility concerned, as well as the physical characteristics, mandatory specification, packaging form, units of measure and standards.

The technical specification, which contains the general characteristics of the requirement item offered in the market, should be prepared in a way that will provide competition except that it is a product that can be obtained from a single source. The technical specifications must be prepared by qualified persons who will use the product/product. The technical specification should include elements that ensure quality and competition. The technical specification should be simple and understandable. Identify only the product requested.

Third step: Purchasing Methods

- Open Tender Procedure[9]
- Negotiation Procedure
- Direct Supply Procedure[10,11].

Fourth step: Purchasing Processes
The demand and technical specifications established by the clinic management and approved by the health facility are procured to be procured by determining the Purchasing method. The determined needs are determined and tendered by the spending

[9]Procurement procedures for general procurement.
(a) Open Tender Procedure: It is the procedure in which the procurement transactions, whose cost value is determined as TL … and above, are realized.
(b) Negotiation Procedure: It is the procedure in which the purchase transactions are determined with a cost between … TL and … TL.
(c) Direct Procurement Procedure: It is the procedure in which large procurement transactions are realized with a cost of approximately … and less (is limited 12.000 USD or lower in big and 3800 USD small cities for governmental hospitals).
http://www.hukuknediyor.com/2019-yili-dogrudan-temin-limiti/.
(d) Leasing: It is a method of renting the product or service in cases, where it will not be used many times or if the unit cost is very high.

[10]Implementing Regulation on Procurement of Procurement.
http://www.mevzuat.gov.tr/Metin.Aspx?MevzuatKod=7.5.12917&sourceXmlSearch=&MevzuatIliski=0.

[11]Public Procurement Law
http://www.mevzuat.gov.tr/Metin.Aspx?MevzuatKod=1.5.4734&MevzuatIliski=0&sourceXmlSearch=

authority (Secretary-General or Hospital Manager) by determining the procurement procedure.

It is against the legislation that the procurement of the needs is met by bargaining procedure or direct procurement without sufficient justification. Planned procurement and timely needs allow for open tendering in a competitive manner (Article 4734/19).

In addition, the products or devices that are planned to be purchased by the direct procurement method can be taken within the scope of direct procurement limits determined for each year and by the direct procurement method when necessary conditions are met. When a planned need is approved by the health facility managers and included in the procurement process, the conclusion of the process and supplying the relevant need to the clinic requires a minimum of 3 months in the open tender procedure. It is extremely important that the clinical management, who will plan the need, know this process, and make timely needs planning.

Fifth step: Establishment of Tender Commission

Article 5 of the "Public Procurement General Communiqué" stating the principles to be clarified and clarifying the hesitations encountered in the implementation of the Public Procurement Law No. 4734 and the Public Procurement Contracts Law. Accordingly, with the consent of the authorizing officer: 1 Commission President, 1 Financial Member, 1 Members and at least 2 Expert Members, in total, at least 5 members shall be appointed together with their reserves.

Sixth step: Goods Acceptance and Inspection

Procurement of goods/devices that are purchased by the purchasing units and contracted are brought to the goods acceptance stage by ordering by the inventory units. In the process of Acceptance of Goods, the products are accepted by carrying out the process of inspection and inspection in accordance with the technical specifications and also within the scope of the Regulation on Inspection and Acceptance of Goods Procurement.

The most important stage of the purchasing process is the inspection stage. During the inspection phase, the commission examines the product, which is decided to be purchased by tender, according to the technical specification in the tender document and concludes the process (Purchasing Principles for Turkish Otorhinolaryngology Head and Neck Surgery Foundation).

10.4 Fraud Cases About Procurement from Turkish Newspapers

Case 1

In the operation organized by Bursa Police, 2 out of 19 people who were detained on charges of harming the state by damaging the state in Orhangazi District State Hospital for misleading tenders were sent to prison and 15 of the suspects, including Yalova State Hospital Chief Physician, was put out of the ban. According to information obtained, Orhangazi District State Hospital in the tenders of irregularities in the state of the damage caused by the claim that among the hospital chief physician of the period and still Yalova State Hospital Chief Physician Op. Dr. 19 people, including Adam C., Bursa Police Department Smuggling and Organized Crime Branch Financial Bureau Chiefs were detained by teams. Suspects completed in police interrogations, 'tender to mislead the state by damaging the allegations' was referred to the court. One of the suspects who testified to the public prosecutor medical company officials Ahmet A. and Kuddüsü C. and Orhangazi Hospital Deputy Director Neşe D. was released, the other 16 suspects were sent to the court on duty to arrest. Hospital purchasing officers Kadir O. and Ismet S. were arrested and sent to prison. 15 people including chief physician Adem Ç. were subjected to judicial control and a ban on going abroad. It was claimed that the suspects incurred mischief in the tender and caused the state to suffer 77 thousand Turkish Liras (Cihan Haber, Devlet Hastanesindeki Yolsuzluk Operasyonunda 2 Kişi Tutuklandı). https://www.haberler.com/devlet-hastanesindeki-yolsuzluk-operasyonunda-2-haberi/ 02.09.2008.

Case 2
According to allegations, Gazi Yaşargil Training and Research Hospital, a Deputy Director in charge of Purchasing Commission last year, 12-month Air Conditioning Maintenance and Repairs Hospital Children's Hospital by changing the tender specifications, a company close to him that the tender of 192 thousand TL 40 thousand TL by the management of the Children's Hospital for another year revealed unfair earnings. Again in 2 years, 2 care was needed, and the removal of 12 care was reacted. According to the allegations, now the same Deputy Director, Gazi Yaşargil Training and Research Hospital's tender for the materials prepared for the Chiller Cooling Group, the same company has been prepared to receive the appropriate specification according to him. With the specification to be prepared, only the said company will be awarded the tender. According to the specification prepared according to the company, anticompetitive substances are put in place and unfair profit is tried to be obtained.

10.4.1 Compressor Maintenance for the Same Company

Allegedly, it was learned that the compressor maintenance repair specification for the same company was prepared. It was learned that the ISO 9001 certificate was requested from the company. It was reported that this document was an ISO document in the administrative sense and was not related to maintenance repairs. Alarko, which is an importer company and sells only products, cannot be repaired by the compressor.

10.4.2 Specifications for Single Companies Fly in the Air

Gazi Yaşargil Training and Research Hospital, 22 A event after the outbreak of this time the specifications on the questioners stop, do not know, specifications for single companies are flying in the air. Where is the competitive environment for each firm's biddability, regardless of public interest?

10.4.3 *What Does the Technical Specification Say?*

Article 12 of the Public Procurement Law No: 4734 and Article 19 of the Implementing Regulation are as follows:

"It does not contain any anti-competitive issues. It is obligatory to provide equal opportunity for all tenderers". https://www.diyarbakirgazete.com/yine-arastirma-hastanesi-yine-yolsuzluk/ 9 Nisan 2018.

10.5 Discussion

Action in enterprises always starts with the preparation of the sales budget. Thereafter the purchasing department carries out its budget. This helps us to decide how much and which goods and services we will receive. This stage is a very parameter and very vulnerable stage for the procurement department. This stage is very vulnerable to abuse. If the company does not have written procurement policies and internal control procedures, then abuse becomes even stronger.

According to auditors who have worked in resource-constrained hospitals, three types of fraud are particularly common. These include

(1) diversion of patient fee revenue at the point of service;
(2) diversion of accounts receivable, or checks submitted by patients or companies to pay debts owed on their accounts; and
(3) collusion between hospital purchasing agents and suppliers (Musauc, 2008).

There are many ways to prevent fraud in hospitals. Among many ways, "Fraud examination" and "business ethics" education is the best way to prevent fraud in the health sector. In particular, those who are a student in management school, business school, or health management school have the potential to be senior, executive managers in the health sector should receive very detailed basic and applied continuous education on fraud prevention. Turkey' s internal auditors in the ministry of health must undergo such training. Mobility in enterprises starts with the preparation of the sales budget. Then the purchasing department carries out its works. Budgeting procedure and purchasing procedure should be followed by internal auditor and external auditor in the hospitals. External auditor have responsibilities to control internal auditors' activities according to International Standard on Auditing Using The Work of Internal Auditors. If the company does not have an internal audit department and written internal control procedures do not work, it is difficult to prevent fraud in the purchasing department.

Apart from that, the independent auditing application in public health enterprises can be applied in governmental health institutions.

In the health sector, the establishment of a well-equipped team[12] on corruption prevention is extremely important. In the case of corruption, this team must intervene.

IFRS usage, or proper implementation of the COSO internal control procedures, as well as follow-up of corruption arrested in Turkey, will prevent corruption in the public sector purchasing a change in procurement law. In the procurement law, the obligation of the lowest bidder to win leads to many corruption.

As I mentioned in the Introduction section, it is very important to establish a compliance department that can follow up on the issues of internal auditing, sustainability, business ethics, corporate reputation, efficiency, environmental reporting in health enterprises at the same time under an umbrella. Seven elements are essential for the compliance department (Baumann, 2007).[13]

- Implementing written policies and procedures,
- Designating a compliance officer and compliance committee,
- Conducting effective training and education,
- Developing effective lines of communication,
- Conducting internal monitoring and auditing,
- Enforcing standards through well-publicized disciplinary guides and
- Responding promptly to detected problems and undertaking corrective actions.[14] (OIG, 2003).

As auditors, we know that business abuse will never end. The important thing should always be to avoid proactive methods. For this purpose, Fraud Simulation Tool,[15] artificial intelligence, and game theory have been developed.

---

[12]The First Medicare Fraud Strike Force (Strike Force) was launched in March 2007 as part of the South Florida Initiative a joint investigative and prosecutorial effort against Medicare fraud and abuse among Durable Medical Equipment (DME) suppliers and Human Immunodeficiency Virus (HIV) infusion therapy providers in South Florida. The Strike Force teams use advanced data analysis techniques to identify high-billing levels in health care fraud hot spots so that interagency teams can target emerging and migrating schemes along with chronic fraud by criminals masquerading as health care providers and suppliers (Townsend, 2013).

[13]Model Corporate Compliance Programs: OIG has published compliance program guidance for 11 segments of the health care industry: hospitals, clinic laboratories, durable medical equipment, third-party equipment, prosthetics and supplies providers, nursing facilities, hospices, medicare + choice.

[14]Office of Inspector General's Compliance Program Guidance for Pharmaceutical Manufacturers, OIG Compliance Program Guidance for Pharmaceutical Manufacturers, 68 Fed. Reg.23,731, https://oig.hhs.gov/fraud/docs/complianceguidance/042803pharmacymfgnonfr.pdf, (May 5, 2003).

[15]Using the RetSim Fraud Simulation Tool to Set Thresholds for Triage of Retail Fraud, October 2015, https://doi.org/10.1007/978-3-319-26502-5, Conference: The 20th Nordic Conference on Secure IT Systems.

References

ACFE. (2018). *Report to the nations 2018 global study on occupational fraud and abuse.* Retrieved March 14, 2019, https://s3-us-west-2.amazonaws.com/acfepublic/2018-report-to-the-nations.pdf.

Atkinson, R. D., Ezell, S. J., Andes, S. M., Castro, D. D., & Bennett, R. (2010). The internet economy 25 years after.com: transforming life and commerce. *Information Technology and Innovation Foundation (ITIF).* Retrieved March 3, 2019, from www.itif.org/files/2010-25-years.pdf.

Bardhan, P. (1997). Corruption and development: A review of issues. *Journal of Economic Literature, 35*(3), 1320–1346.

Baumann, L. (2007) Corporate compliance programme. In L. Baumann, T. S. Crane, P. Hooper, D. M. Barry, B. H. Ryland, A. Humphreys, et al. (Eds.), *Health Care Fraud and Abuse: Practical Perspectives* (2nd ed.). Arlington, VA: American Bar Association, ABA Health Law Association, BNA Books.

Bozkus Kahyaoglu, S., & Caliyurt, K. (2018). Cyber security assurance process from the internal audit perspective. *Managerial Auditing Journal, 33*(4), 360–376.

Caliyurt, K., Agun, C., & Pamukcu, A. (2005). Recent developments on internal auditing legislation in public institutions of Turkey: A hospital case, in *Third European Academic Conference 'Internal Audit and Corporate Governance at the Cass Business School', London, England 2005.*

Carter, P. L., & Monczka, R. M. (1978). MRO inventory pooling? *International Journal Purchasing Supply Management, 14,* 27.

Carter, J. R., & Narasimhan, R. (1996). Is purchasing really strategic? *International Journal Purchasing Supply Management, 32,* 20–28. https://doi.org/10.1111/j.1745-493X.1996.tb00216.x.

Cohen-Kohler, J. (2006). Pharmaceuticals and corruption: A risk assessment. In J. Kotalik, D. Rodriguez (Eds.), *Global corruption report 2006.* Transparency International (pp. 77–85). London: Pluto Press. Retrieved March 8, 2019, from http://www1.worldbank.org/publicsector/anticorrupt/corecourse2007/Pharmaceuticals.pdf.

Datapine. (2019). Procurement key performance indicators and metrics. Retrieved March 3, 2019, from https://www.datapine.com/kpi-examples-and-templates/procurement.

Dawson, S., & Babinchak, J. (2019). COSO issues guidance for healthcare providers collaboration with crowe, common spirit health aims to strengthen industry governance and internal control. Retrieved May 15, 2019, from https://www.coso.org/Documents/COSO-Issues-Guidance-for-Healthcare-Providers.pdf.

Deloitte. (2019). *A roadmap to applying the new leasing standard.* Retrieved May 5, 2019, from https://www.iasplus.com/en/publications/us/roadmap-series/leases.

Diyarbakır Newspaper. (2018). *Yine Araştırma Hastanesi, Yine yolsuzluk*! https://www.diyarbakirgazete.com/yine-arastirma-hastanesi-yine-yolsuzluk/ 9 Nisan 2018.

Ferilli, A. (2013). *Fraud detection and deterrence in medicare and medicaid.* New York: Nova Publishers.

Ford, E. W., & Scanlon, D. P. (2007). Promise and problems with supply chain management approaches to health care purchasing. *Health Care Management Review, 32,* 192–202.

GAO. (2018, July 17). Testimony before the subcommittee on oversight, Committee on ways and means, house of representatives, United States Government Accountability Office MEDICARE actions needed to better manage fraud risks. https://www.gao.gov/assets/700/693156.pdf.

González-Benito, J. (2007). A theory of purchasing's contribution to business performance. *Journal of Operation Management, 25,* 901–917. https://doi.org/10.1016/j.jom.2007.02.001.

Haberler Newspaper. (2019). *Devlet Hastanesindeki Yolsuzluk Operasyonunda 2 Kişi Tutuklandı.* Retrieved February 13, 2019, from https://www.haberler.com/devlet-hastanesindeki-yolsuzluk-operasyonunda-2-haberi/.

Hadad, S., Hadad, Y., & Simon-Tuval, T. (2013). Determinants of healthcare system's efficiency in OECD countries. *The European Journal of Health Economics, 14,* 253–265.

IFRS. (2019a). Why global accounting standards? Retrieved May 19, 2019, from https://www.ifrs.org/use-around-the-world/why-global-accounting-standards/.

IFRS. (2019b). Leases. Retrieved May 19, 2019, from https://www.ifrs.org/issued-standards/list-of-standards/ifrs-16-leases/.

IIA. (2019). About internal auditing. Retrieved May 16, 2019, from https://global.theiia.org/about/about-internal-auditing/Pages/About-Internal-Auditing.aspx.

Investopedia. (2019a). What is an asset? Retrieved May 19, 2019, from https://www.investopedia.com/terms/a/asset.asp.

Investopedia. (2019b). Fixed asset. Retrieved May 19, 2019, from https://www.investopedia.com/terms/f/fixedasset.asp.

Liu, X. (2018). *Discussion on internal accounting control in hospital*. Crimson Publisher. Retrieved January 6, 2018, from https://crimsonpublishers.com/downloadPdf.php?folder=rmes&file=RMES.000563.

Mackey, T. K., & Liang, B. A. (2012). Combating healthcare corruption and fraud with improved global health governance. *BMC International Health and Human Rights, 12*(1).

Maruthappu, M., & Keogh, B. (2015). The future of health system leadership. *Lancet, 385*(2325). Retrieved February 23, 2019, from https://doi.org/10.1016/s0140-6736(15)60990-4.

Musauc, S. (2008). *Fraud in hospitals* (4 pp.). Bergen: Chr. Michelsen Institute (U4 Brief 2008:8). Retrieved March, 15, 2019, from https://www.u4.no/publications/fraud-in-hospitals.

Naylor, C. D. (1999). Health care in Canada: Incrementalism under fiscal duress. *Health Affairs, 18,* 9–26.

Nextenders. (2016). 6 types of procurement fraud and how to spot them. Retrieved January 17, 2019, from http://www.nextenders.co.uk/6-types-of-procurement-fraud-and-how-to-spot-them/.

Nollet, J., Beaulieu, M., & Fabbe-Costes, N. (2017). The impact of performance measurement on purchasing group dynamics: The Canadian experience. *Journal of Purchasing and Supply Management, 23*(1), 17–27. https://doi.org/10.1016/j.pursup.2016.04.001.

Office of Inspector General's Compliance Program Guidance for Pharmaceutical Manufacturers. (2003, May 5). OIG Compliance Program Guidance for Pharmautical Manufacturers, 68 Fed. Reg.23,731. Retrieved March 4, 2019, from https://oig.hhs.gov/fraud/docs/complianceguidance/042803pharmacymfgnonfr.pdf.

Özkul, F.U., & Pamukçu, F. (2012). Fraud detection and forensic accounting. In K. Caliyurt & S. Idowu (Eds.), *Emerging fraud: Fraud cases from emerging economies*. Berlin Heidelberg: Springer.

Paschke, A., Dimancesco, D., Vian, T., Kohler, J., & Forte, G. (2018). Increasing transparency and accountability in national pharmaceutical systems. Retrieved February 16, 2019, from https://www.researchgate.net/publication/328634442_Increasing_transparency_and_accountability_in_national_pharmaceutical_systems.

Pooler, V. H., Pooler, D. J., & ve Farney, S. D. (2004). *Global purchasing and supply management fulfill the vision* (2nd ed.). Boston: Kluwer Academic Publishers.

Reck, R. F., & Long, B. (1988). Purchasing: A competitive weapon. *Journal of Purchasing and Materials Management, 24,* 2–8.

Republic of Turkey, Ministry of Health. (2019). *General Directorate of Public Hospitals*, General Health Statistics. Retrieved March 3, 2019, from https://rapor.saglik.gov.tr/istatistik/rapor/.

Roehrig, C. S. (2011). Will the heath care cost curve be bent? Where we stand at the start of 2011. *Business Economics, 46,* 159–162.

Schandl, A., & Foster, P. (2019, January). COSO internal control—Integrated framework: An implementation guide for the healthcare provider industry. Retrieved May 5, 2019, from https://www.coso.org/Documents/COSO-CROWE-COSO-Internal-Control-Integrated-Framework.pdf.

Scheuing, D. R. (1984). Size exclusion chromatography of polyelectrolytes in dimethylformamide. https://doi.org/10.1002/app.1984.070290912. Retrieved January 10, 2019, from https://onlinelibrary.wiley.com/doi/abs/10.1002/app.1984.070290912.

Schütz, K., Kässer, M., Blome, C., & Foerstl, K. (2019). How to achieve cost savings and strategic performance in purchasing simultaneously: A knowledge-based view. *Journal of Purchasing and Supply Management*. https://doi.org/10.1016/j.pursup.2019.04.002.

Steinbart, P. J., Raschke, R. L., Gal, G., & Dilla, W. N. (2012). The relationship between internal audit and information security: An exploratory investigation. *International Journal of Accounting Information Systems, 13*(3), 228–243.

Steinbart, P. J., Raschke, R. L., Gal, G., & Dilla, W. N. (2013). Information security professionals' perceptions about the relationship between the information security and internal audit functions. *Journal of Information Systems: Fall 2013, 27*(2), 65–86.

Steinbart, P. J., Raschke, R., Gal, G., & Dilla, W. N. (2015). The influence of internal audit on information security effectiveness: Perceptions of internal auditors. https://papers.ssrn.com/sol3/papers.cfm?abstract_id=2685943.

SDG16. Sustainable development goal 16. https://sustainabledevelopment.un.org/sdg16.

The World Health, Report. (2010). Health systems financing: The path to universal coverage. Geneva: World Health Organization. Retrieved June 1, 2019, from http://www.who.int/whr/2010/en/.

Townsend, L. (2013). *The health care fraud and abuse program*. New York: Nova Publishers.

Türk Kulak Burun Boğaz ve Baş Boyun Cerrahisi Vakfı *Satın Alma Prosedürleri* (Purchasing Principles for Turkish Otorhinolaryngology Head and Neck Surgery Foundation). Retrieved May 6, 2019, from http://www.tkbbv.org.tr/menu/11/satin-alma-prosedurleri.

Vian, T. (2008). Review of corruption in the health sector: Theory, methods and interventions. *Health Policy Plan, 23*(2), 83–94.

Vian, T., Savedoff, W. D., & Mathisen H. (2010). *Anticorruption in the health sector: Strategies for transparency and accountability*. Kumarian Press, Sterling, USA.

Wallace, L., Lin, H., & Cefaratti, M. A. (2011). Information security and Sarbanes–Oxley compliance: An exploratory study. *Journal of Information System, 25*, 185–212. Retrieved May 15, 2019, from https://acg6415.wikispaces.com/file/view/JournalþofþInformationþSystemsþ2011þWallace.pdf.

Kıymet Tunca Çalıyurt Prof. Dr. Kiymet Tunca Caliyurt, CPA, CFE, graduated from the Faculty of Business Administration at Marmara University, Istanbul, Turkey. Her Masters and Ph.D degrees are in Accounting and Finance Programme from the Social Graduate School, Marmara University. She has worked as auditor in Horwath Auditing Company, manager in McDonalds and finance staff in Singapore Airlines. After vast experience in private sector, she has started to work in academia. She is holding CFE, CPE and external auditor titles. Her research interests are in accounting, auditing, fraud, social responsibility, corporate governance, finance and business ethics, with a special interest in aviation management, NGOs, women rights in business. She has been as visiting researcher in Massachussetts University Amherst Business School. She is the founder of the International Group on Governance, Fraud, Ethics and Social Responsibility (IGonGFE&SR) which was founded in 2009. In 2009, she also founded the International Women and Business Group, which organizes a global, annual conferences. Kiymet has published papers, book chapters and books both nationally and internationally on fraud, social responsibility, ethics in accounting/finance/aviation disciplines in Springer and Routledge. She is book series editor in Springer with the title Accounting, Finance, Sustainability, Governance & Fraud: Theory and Application, and book series editor in Routledge with the title Women and Sustainable Business. Some book titles: Emerging Fraud (with Sam Idowu), Corporate Governance: An International Perspective (with Sam Idowu), Women and Sustainability in Business: A Global Perspective, Sustainability and Management: An International Perspective (with Ulku Yuksel), Globalization and Social Responsibility (with David Crowther), Regulations and Applications of Ethics in

Business Practice (with Dr. Jiang Bian), Ethics and Sustainability in Accounting and Finance, Volume I. She is acting as member in editorial board Journal of Financial Crime, Social Reponsibility Journal, International Journal on Law and Management, Journal of Money Laundering Control. She is regular speaker at International Economic Crime Symposium in Jesus College, Cambridge University. She is partner of Herme Consulting in Trakya University Technopark.

Chapter 11
Communicating Integrity Framework Information on Malaysian Private Hospitals' Websites

Roshima Said and Corina Joseph

Abstract In ensuring accountability and transparency, an integrity system is vital within an organization, Website can be used to share the integrity initiatives information. The objective of the paper is to determine the extent of Integrity Framework information disclosure on Malaysian private hospital websites. The data collected in this study were based on a content analysis of inclusion or exclusion of integrity framework information disclosed on Malaysian private sector websites and measured using the Private Hospitals-Integrity Framework Disclosure Index (PH-IFDi). The mean for the disclosure level of Private Hospitals-Integrity Framework Disclosure Index (PH-IFDi) is 10.28%. Overall, the average disclosure is below than 60% and relatively low. The Award category rank first followed by Patient's right and responsibilities, Mission, Vision and core values category. From the findings of the study, it shows that the Healthcare disclosed more items that portray the image of integrity in a way to preserve the image and reputation of the organization. this study has brought up a number of implications that may be applied in many countries, whether they are developing or developed countries. This study significantly provided useful information in helping the regulatory authorities and policy makers to strengthen their mechanisms to uphold integrity among health care provider. This study also has significant implications to the development of policy and guidelines for the assessment of Private Hospitals-Integrity Framework Disclosure Index (PH-IFDi).

Keywords Integrity · Accountability · Websites · Healthcare · Content analysis

---

R. Said (✉) · C. Joseph
Faculty of Accountancy, Universiti Teknologi Mara, Shah Alam, Malaysia
e-mail: roshima712@uitm.edu.my

C. Joseph
e-mail: corina@uitm.edu.my

K. T. Çalıyurt (ed.), *Integrity, Transparency and Corruption in Healthcare & Research on Health, Volume I*, Accounting, Finance, Sustainability, Governance & Fraud: Theory and Application, https://doi.org/10.1007/978-981-15-1424-1_11

11.1 Introduction

Transparency International (2016) defined integrity as 'behaviours and actions consistent with a set of moral or ethical principles and standards, embraced by individuals as well as institutions, that create a barrier to corruption'. Integrity remains popular in many research fields, for example, ethics, management, and psychology. Integrity is a basis of good governance and many countries have emphasized the integrity management (OECD, 2009). The Integrity Framework is a systemic and comprehensive approach that consists of instruments, processes, and structures for advancing integrity and combatting corruption in government agencies, which can also be applied to private sectors. OECD (2009) classified the key components of an Integrity Framework into three main pillars: instruments, processes (planning, implementing, evaluating and adapting) and structures (i.e., the organizational aspect of integrity management.

In ensuring accountability and transparency, an integrity system is vital within an organization, good corporate governance plays a key role in maintaining corporate integrity and managing the risk of unethical and unlawful activities (Agyei-Mensah, 2017). Being transparent is essential to improve the governance system. In the private sector, to achieve transparency, disclosure of information could enable the evaluation by investors on organizations' allocation of resource management (Agyei-Mensah, 2017). Disclosure is a way to improve communication and awareness regarding the integrity framework. In addition, disclosure of the organization's policies such as integrity framework is an essential element in achieving good governance (Midin, Joseph, Mohamed, 2017).

Corruption is one of the negative effects of lacking integrity. The Malaysian government has undertaken several efforts, to combat corruption and promote integrity in society. Malaysia is selected as a case study because it is aiming to become a fully developed nation in 2020. Good governance is one of the preconditions to achieve high-income nation status. Since the 1980s, among the major institutions set up are the Anti-Corruption Agency, the Public Accounts Committee and the Malaysian Institute of Integrity (Siddique, 2010). However, these formal measures are blamed to be inefficient in reducing unethical transactions and wrongdoings by society (Siddique, 2010). For this reason, Joseph, Janggu, Rahmat, Madi, and Mohamed (2017) suggested that all these institutional mechanisms need to be reassessed.

This paper focusses on health care because it has unique characteristics. In health care, customers are the immediate patients followed by their families and quite possibly their friends, as the outcome of the healthcare service potentially affects all their lives (Ahmed, Abd Manaf, & Islam, 2017). The healthcare sector is facing more pressures nowadays compared to the past. The causes are: increase in healthcare expenditure, stiff competition, advancement in technology, development of new illness and medication, people's consciousness, rise in customers' complaints and media's attention. On the other hand, the need for better services, medications, and curative discoveries are offset by a decrease amount in investment capital and scarcity of volunteers and qualified caregivers (Jamali, Hallal, & Abdallah, 2010). Therefore, private hospitals must ensure all the business operations are conducted in an orderly and ethical manner. Private hospitals in the Western countries have published a code of conduct. For example, the Stanford University Medical Center, United States, implements a comprehensive code of conduct comprising

1. Integrity in patient care;
2. Integrity billing and financial matters;
3. Integrity in business conduct;
4. Integrity with referral sources;
5. Integrity in the workplace;
6. Integrity in decision-marking;
   (Stanford Health Care, n.d.)

Another example of the code of business conduct is by Carolinas HealthCare System. Carolinas HealthCare System identified four core values: commitment, caring, teamwork, and integrity. The published booklet "A System of Integrity' includes standards in which personal and professional behaviors that need to be followed. The standards include: (1) Patient Care; (2) Privacy and Confidentiality; (3) Conflicts of Interest; (4) Business Courtesies, Gifts and Customer–Supplier Relations; (5) Competition and Antitrust; (6) Safety, Health, and Environment; (7) Safeguarding System Property and Technology; (8) Proper Accounting and Record-Keeping; (9) Billing and Coding Integrity; (10) Identifying and Reporting False Claims; (11) Admission, Treatments, and Referrals; (12) Community Involvement and Political Activities; (13) Dealing with Government Officials and Regulatory Agencies; (14) Marketing and Media Inquiries; (15) Getting Help; (16) Guidance for Business Partners, and (17) Compliance Education and Training. This signifies that private hospitals in the western countries are practicing transparency on the organization's ethical conduct.

In Malaysia, the Academy of Medicine has introduced the Ethical Professional Practice Guideline, which is, intended to provide guidance to medical practitioners on matters related to ethics, professional conduct and responsibilities to patients. Professional practice is now subjected to greater scrutiny as a result to more open discussions on standards by the public and media (Teoh, 2017). It was asserted that trust from the patient is important in the professional relationship. Therefore, the confidentiality of patient's information and the management of conflicts need to be prioritized. The conflict with patients can be resolved by having clear communication, which, in turn, can prevent patients from initiating legal actions (Academy of Medicine of Malaysia, n.d.).

All internal and external members of organizations need to be informed about integrity initiatives, for example, anti-corruption practices (ACP) (Coonjohn & Lodin, 2011; Joseph et al., 2016). The website can be used to share the integrity initiatives information. It is maintained in this paper that the integrity framework information disclosures on the private hospital websites are necessary for enhancing good governance. Therefore, the following research questions is: What is the extent of integrity framework information disclosed on private hospital websites?

The corresponding objective of the paper is to determine the extent of the Integrity Framework information disclosure on Malaysian private hospital websites. This paper is significant for the following reasons—there is little published research examining the level of integrity framework disclosure in developing countries such as Malaysia. Previous studies on integrity framework disclosure focussed on local authorities (for example, Joseph et al., 2017). In addition, the websites of private hospitals were analyzed to explain the promotion of medical tourism (Moghavvemi et al., 2017). The findings would assist in helping the regulatory authorities and policymakers to strengthen their mechanisms to uphold integrity. The remainder of this paper proceeds as follows. Section 11.2 presents the literature review. In Sect. 11.3, the methodology is explained. Sections 11.4 and 11.5 provide the analysis of results and discussions, respectively. Concluding comments are presented in Sect. 11.6.

11.2 Literature Review

The organization's website is a strategic communication tool to share information about corruption. Gordon and Wynhoven (2003) examined the top 100 nonfinancial enterprises' websites to comprehend on corruption, management, and reporting practices. The anti-corruption information was disclosed by 43% firms' websites. Transparency International (2009) analyzed the ACP disclosure by the Forbes Global 250 companies related to overall organizational practices. It was concluded that firms frequently report high-level strategic commitments to anti-corruption efforts. Hess (2009) stated that the ACP disclosure by companies is helping in: achieving the organization's goals, increasing public awareness, enhancing accountability and promoting learning organization. Moreover, by disclosing ACP, it indicates the firms' knowledge on the adverse effects for their reputation and respectability by society, as a result of involving in unethical and unlawful transactions, for example, corruption (ACCA, 2008).

ACCA (2008) evaluated the execution, disclosure, and anti-corruption mechanisms on the top 50 Australian public firms and discovered that most of the firms' disclosure bribery and corruption has frequently been shallow and inadequate. Only a small number of ASX50 companies disclosed on the countering bribery and corruption in detail. It was found that most disclosures are incomplete—inadequate public communication on anti-corruption programs and lacking anti-corruption procedures.

Joseph et al. (2016) carried out a study to compare the extent of anti-corruption information disclosure in annual reports for Malaysia and Indonesia companies. Results show that anti-corruption practices (ACP) in both countries are still in their infancy stages. The anti-corruption codes of conduct and whistle-blowing practices were the most reported item (Joseph et al., 2016).

It has been highlighted that integrity could be enhanced using the technology evolution (Said, Alam, & Abd Aziz, 2015), for example, website. In the private sector, communicating on anti-corruption strategies to the public complement a company's actual engagement in anti-corruption initiatives. The disclosure can increase awareness of anti-corruption measures among stakeholders, and sequentially gives credibility to a company's anti-corruption engagement, as it permits its stakeholder to scrutinize the company's initiatives. It was asserted that the reporting on the firm's engagement with the corruption process can be explored and examined by using the institutional theory (Barkemeyera, Preussb, & Lee, 2015).

11.3 Research Methodology

The data collected in this study were based on a content analysis of inclusion or exclusion of integrity framework information disclosed on Malaysian private sector websites. Content analysis has long been applied in disclosure studies; therefore, it is still relevant in measuring the degree of integrity framework disclosure. This paper develops the Private Hospitals-Integrity Framework Disclosure Index (PH-IFDi) based on the following:

1. The Integrity Framework Disclosure Index (IFDi) developed by Joseph et al. (2017) based on Malaysian Local Authority websites.
2. The Malaysian Society for Quality in Health (MSQH)—a not for profit organization's websites (mission, vision, and values).
3. Sample of real disclosures by three websites (IJN, Normah Medical Specialist Centre and Sime Darby Medical Centre).
4. Sample of Code of Conduct of Stanford University Medical Center and Carolinas HealthCare System.

The final PH-IFDI instrument consists of 8 categories and 56 items. The categories include

(1) Mission;
(2) Vision;
(3) Patient's right and responsibilities;
(4) Objectives;
(5) Core values;
(6) Principles;
(7) Governance; and
(8) Award (see Table 11.1).

The sample selection of this research is 41 private hospital websites. The list of private hospitals is available from the Association of Private Hospitals of Malaysia websites: (http://www.hospitals-malaysia.org/portal/index.asp?menuid=42).

Due to the rapid changes in websites, the data collection is only confined to one month, i.e., February 2017. The data is analyzed using the Statistical Package for Social Science (SPSS).

Table 11.1 Total items for private hospitals-integrity framework disclosure index (PH-IFDi)

Categories	No. of items
Mission	7
Vision	4
Patient's right and responsibilities	2
Objectives	12
Core values	16
Principles	5
Governance	9
Award	1
Total	56

11.4 Findings of the Study

Table 11.2 showed the descriptive statistics for the Private Hospitals-Integrity Framework Disclosure Index (PH-IFDi) from the websites for the year 2017. The descriptive statistics include statistics such as minimum, maximum and mean for each of the eight (8) categories. The mean for the disclosure level of Private Hospitals-Integrity Framework Disclosure Index (PH-IFDi) is 10.28%.

Table 11.3 showed the descriptive statistics for the disclosure of Private Hospitals-

Table 11.2 Descriptive statistics for private hospitals-integrity framework disclosure index (PH-IFDi)

Private hospitals-integrity framework disclosure index (PH-IFDi)	%
Mean	10.28
Minimum	0.00
Maximum	28.57

Table 11.3 Descriptive statistics for private hospitals-integrity framework disclosure index (PH-IFDi)—by categories

	Mean	Minimum	Maximum
Mission	14.76	0	42.86
Vision	14.17	0	75.00
Patient's right and responsibilities	21.67	0	100.00
Objectives	6.67	0	58.33
Core values	8.96	0	62.50
Principles	3.33	0	100.00
Governance	8.15	0	33.33
Award	60.00	0	100.00

Table 11.4 Descriptive statistics for private hospitals-integrity framework disclosure index (PH-IFDi)—rank by themes

	Mean	Rank
Mission	14.76	3
Vision	14.17	4
Patient's right and responsibilities	21.67	2
Objectives	6.67	7
Core values	8.96	5
Principles	3.33	8
Governance	8.15	6
Award	60.00	1

Integrity Framework Disclosure Index (PH-IFDi) from the company's websites. The descriptive statistics include statistics such as minimum, maximum and mean for each of the eight (8) categories of Private Hospitals-Integrity Framework Disclosure Index (PH-IFDi). The mean for the disclosure level of Mission category, is 14.76%, whereas for Vision Category, Patient's right and responsibilities category, Objectives category, Core Values category, Principles category, Governance category and Award category is 14.17%, 21.67%, 6.67%, 8.96%, 3.33%, 8.15%, and 60% respectively. Overall, the average disclosure is below 60% and relatively low.

Table 11.4 shows the ranking category of the Private Hospitals-Integrity Framework Disclosure Index (PH-IFDi). The Award category rank first followed by Patient's right and responsibilities, Mission, Vision, and core values category.

11.5 Discussion and Conclusion

Communicating identity through online system (i.e. website) is important as website reporting has many roles which include; performance tracking, reputation building, and assessment of activities (Fatma & Rahman, 2014). In addition, for several years online system development has changed and affected people's lives in many ways, especially in relation to process, exchange, and distribution of digital information. Online-enabled communication has improved reach, efficiency and accuracy, at work and in personal lives. It gives users quick access to data, information and knowledge at a very low cost and from a wide range of people, communities and cultures through Information Management System (IMS) capabilities. Clear communication and information were a criterion for the public's perception of satisfaction with the healthcare services. The public emphasized the importance of receiving adequate information and there was a need for the information they received to be improved for their satisfaction to increase.

The findings of the study show that the award category score the highest rank of the Private Hospitals-Integrity Framework Disclosure Index (PH-IFDi). The reason to disclose on the award-winning is as a recognition to attract and retain clients/ patients and to be publicly recognized around the globe as a great health provider. Anas, Abdul Rashid, and Annuar (2015) investigates the effect of the award on CSR disclosures in annual reports of Malaysian PLCs and found a statistically significant result. Past studies had shown, there are various drivers for CSR reporting and environmental reporting. ACCA (2002) emphasized that among the driving forces for environmental reporting in Malaysia are the introduction of the Malaysia Code on Corporate Governance listing requirements, the National Annual Corporate Award (NACRA) and ACCA Award named as Malaysian Environmental Reporting Award (MERA), recently in year 2004 changed to Malaysian Environmental and Social Reporting Award (MESRA).

The second most disclosed category is Patient's rights and responsibilities. The aim is to encourage positive relations between health care and patients and to promote best practices in hospitals. According to Dempski (2008), patients have the right to expect high-quality care such as

1. The right to information,
2. The right to choose,
3. Patients have the right to access emergency health services when and where the needs arises,
4. Patients have the right to fully participate in all decisions that relate to their health care,
5. Care without discrimination,
6. The right to privacy,
7. The right to speedy complaint resolution,
8. Taking on new responsibilities for maintaining good health.

While Scowen (2006) states the following as the patient's responsibilities:

1. To treat the staff who care for them with courtesy and respect at all times.
2. To inform the hospital at once of any change in contact address or telephone number.
3. To attend an appointment given on time, or give reasonable notice of inability to attend.
4. To understand that there are pressures and limitations of resources on the health services and those working within it.
5. To consider the consequences of refusing treatment or not following medical advice and accept responsibility for their own actions.
6. To let the staff know if they have any allergies or sensitivities to medications.
7. To give staff full information about their condition including permanent disabilities, along with details of any medicine they are taking.
8. To tell the staff if they are uncertain or don't understand any aspect of their treatment.
9. To tell the staff about any change in their health that could affect the treatment they are having.

10. To take any medicines as instructed and seek medical advice before stopping or changing any treatment.
11. To treat other patients with courtesy and respect.
12. To attend follow-up appointments as requested.

Furthermore, the findings of the study show that the Mission category scores the third highest rank in the Private Hospitals-Integrity Framework Disclosure Index (PH-IFDi) via websites. Bart and Tabone (1998) in Forehand (2000), define a mission statement as a written formal document that attempts to capture an organization's unique and enduring purpose and practices. While, Rigby (1998) in Forehand (2000) suggests that, a mission statement should do is communicate to what are timidly referred to as the company's key stakeholders what it is that the company is doing and ought to be doing. The objectives, mission, and vision are crucial to management because they are components of the three broad goals of management (King, 1994):

1. To create vision and mission statements that clearly state the purpose of the organization;
2. To communicate these statements in a succinct and effective manner; and
3. To provide a means to measure and encourage future progress.

Additionally, Bart and Tabone (1998) identified eight "performance-enhancing drivers" that should be included in the mission statement, that affected the employee's behavior. The eight drivers identified as significant were

1. To provide a sense of purpose;
2. To increase CEO control;
3. To define behavior standards;
4. To enable employees to identify with their organization;
5. To give greater recognition to the interests of external stakeholders;
6. To inspire and motivate employees;
7. To refocus the organization during a crisis; and
8. To improve the resource allocation process.

Having honesty and integrity in the workplace is one of the most important qualities in fostering the success of the organization. According to Coughlin, Barker, and Dawson (2012), there are three types of integrity namely internal integrity, external integrity, and the image of integrity. Internal Integrity is the fundamental of a trustworthy person. Internal integrity is referring to "doing what you think is the right thing to do even when no one else is looking". While external integrity is what you show other people. External integrity affects not only business relationships, but it can also affect the organization's brand or reputation. Upholding high levels of integrity is not only significant for the results for the individual to generate, but it is also important for the results of the organization. Whereas, the image of integrity is referring to what the person or organization does to protect the image of integrity, reputation as a person or organization of integrity. From the findings of the study, it shows that the health care disclosed more items that portray the image of integrity in a way to preserve the image and reputation of the organization.

Integrity and accountability are related and it is essential to ensuring the trust between the public and healthcare professionals. Integrity and accountability form the basis of the "social contract" between healthcare and society, which contributes professionals the privilege of self-regulation. The need for an ethical foundation is conceivably more critical in healthcare situations than in many other environments. As we know that healthcare is an industry that affects the lives of every single person. Patients and family members desire high-quality care, patient safety, and the latest medical technology. But, ultimately, trust in the integrity of the medical professionals is the foundation for their peace of mind. If an organization's first priority is to provide the best possible service to its customers or patients, then integrity and ethical behavior naturally follow. Keeping the welfare of the customer or patient in mind at all times leads to good decisions.

This study, like any other researches, has its own limitations. Among them is the source of data. The sources of data in this study are limited to companies' websites. The study is a cross-sectional study where a data-collection strategy in which data is collected at one point in time and only focuses on secondary data namely companies' websites. Finally, this study has brought up a number of implications that may be applied in many countries, whether they are developing or developed countries. This study significantly provided useful information in helping the regulatory authorities and policymakers to strengthen their mechanisms to uphold integrity among healthcare providers. This study also has significant implications for the development of policy and guidelines for the assessment of the Private Hospitals-Integrity Framework Disclosure Index (PH-IFDi).

References

Academy of Medicine of Malaysia. (n.d.). *Ethical professional practice guideline*. http://webcache.googleusercontent.com/search?q=cache:2A4QZpdHcTQJ:www.acadmed.org.my/view_file.cfm%3Ffileid%3D786+&cd=2&hl=en&ct=clnk.

ACCA. (2008). Reporting trilogy—Research on reporting disclosures part 1: anti-corruption and corruption reporting disclosures.

Agyei-Mensah, B. K. (2017), (2016). The relationship between corporate governance, corruption and forward-looking information disclosure: A comparative study. *Corporate Governance: The International Journal of Business in Society, 17*(2), 284–304.

Ahmed, S., Abd Manaf, N. S., & Islam, R. (2017). Measuring quality performance between public and private hospitals in Malaysia. *International Journal of Quality and Service Sciences, 9*(2), 218–228. https://doi.org/10.1108/IJQSS-02-2017-0015.

Anas, A., Abdul Rashid, H. M., & Annuar, H. A. (2015). The effect of award on CSR disclosures in annual reports of Malaysian PLCs. *Social Responsibility Journal, 11*(4), 831–852.

Barkemeyera, R., Preussb, L., & Lee, L. (2015). Corporate reporting on corruption: An international comparison. *Accounting Forum, 39,* 349–365.

Bart, C. K., & Tabone, J. C. (1998). Mission statement rationales and organizational alignment in the not-for-profit health care sector. *Health Care Management Review, 23*(Fall), 54–70.

Carolinas HealthCare System. (n.d.). *A system of integrity*. https://www.carolinashealthcare.org/documents/CHS/CHS-Corporate-Code-of-Conduct.pdf.

Coonjohn, J. J., & Lodin, A. (2011). *Developing a strategic implementation plan for anti-corruption*. Retrieved from http://www.jjcoonjohn.com/pdf/Strategic_Implementation_Plan.pdf.

Coughlin, S. S., Barker, A., & Dawson, A. (2012). Ethics and scientific integrity in public health, epidemiological and clinical research. *Public Health Reviews, 34*(1), 5.

Dempski, K. (2008). Patients' rights and responsibilities. *Essentials of Nursing Law and Ethics, 62*.

Fatma, M., & Rahman, Z. (2014). Building a corporate identity using corporate social responsibility: A website based study of Indian banks. *Social Responsibility Journal, 10*(4), 591–601.

Forehand, A. (2000). Mission and organizational performance in the healthcare industry. *Journal of Healthcare Management, 45*(4), 267–277.

Gordon, K., & Wynhoven, U. (2003). *Business approaches to combatting corrupt practices* (pp. 127–152). OECD Annual Report on the OCD Guidelines on Multinational Enterprises: Enhancing the Role of Business in the Fight against Corruption, Paris, OECD.

Hess, D. (2009). Catalyzing corporate commitment to combatting corruption. *Journal of Business Ethics, 88*, 781–790.

Jamali, D., Hallal, M., & Abdallah, H. (2010). Corporate governance and corporate social responsibility: Evidence from the healthcare sector. *Corporate Governance: The international journal of business in society, 10*(5), 590–602.

Joseph, C., Gunawan, J., Sawani, Y., Rahmat, M., Noyem, J. A., & Darus, F. (2016). A comparative study of anti-corruption practices disclosure among Malaysian and Indonesian corporate social responsibility (CSR) best practice companies. *Journal of Cleaner Production, 102*, 2896–2906.

Joseph, C., Janggu, T., Rahmat, M., Madi, N., & Mohamed, N. (2017). Development of the integrity framework disclosure index (IFDI) for Malaysian Local authorities. In *A Paper Presented at the International Conference on Business and Social Science (ICoBSS)*, 28 February–1 March, 2017.

King, J. B. (1994). *Business plans to game plans: A practical system for turning strategies into action*. Santa Monica, CA: Merrit Publishing.

Midin, M., Joseph, C., & Mohamed, N. (2017). Promoting societal governance: Stakeholders' engagement disclosure on Malaysian local authorities' websites. *Journal of Cleaner Production, 142*(4), 1672–1683.

Moghavvemi, S., Ormond, M., Musa, G., Mohamed Isa, C. R., Thirumoorthi, T., Mustapha, M. Z., et al. (2017). Connecting with prospective medical tourists online: A cross-sectional analysis of private hospital websites promoting medical tourism in India, Malaysia and Thailand. *Tourism Management, 50*, 154–163.

OECD. (2009). *Public governance and territorial development directorate (GOV)*. Retrieved from: ftp://pogar.org/localuser/pogarp/finances/oecd/oecd-paper.pdf.

Rigby, R. (1998). Mission statements. *Management Today* (March): 56–59.

Said, J., Alam, M. M., & Abd Aziz, M. A. (2015). Public accountability system: Empirical assessment of public sector of Malaysia. *Asian Journal of Scientific Research, 8*(2), 225–236.

Said, J., Alam, M. M., & Khalid, M. A. (2016). Relationship between good governance and integrity system: Empirical study on the public sector of Malaysia. *Humanomics, 32*(2), 151–171.

Scowen, P. (2006). Patients' rights and responsibilities. *The Bulletin of the Royal College of Surgeons of England, 88*(5), 152–154.

Siddique, N. A. (2010). Combatting corruption and managing integrity in Malaysia: A critical overview of recent strategies and initiatives. *Public Organization Review, 10*, 153–171.

Standard Health Care. (n.d.). *Code of conduct*. https://stanfordhealthcare.org/content/dam/SHC/about-us/code-of-onduct/docs/som-code-of-conduct-7-26-11.pdf.

Teoh, S. C. (2017). Continuing development in ethics and professionalism by Malaysian Medical Association. In *Symposium Continuing Development in Ethics and Professionalism*, 229JMAJ, May/June 2007, Vol. 50, No. 3, pp. 228–233.

Transparency International. (2009). *Transparency in reporting on anti-corruption: A report on corporate practices*, Berlin.

Transparency International. (2016). http://www.transparency.org/news/feature/corruption_perceptions_index_2016.

Roshima Said (Ph.D) is an Associate Professor of Universiti Teknologi MARA, Malaysia. Her research interests are in the area of Corporate Governance, Corporate Social Responsibility, Ethics and Financial Criminology. She has conducted a number of researches that were presented at international and national seminars, conferences and published in proceedings and journals. She also has published books and book chapters both nationally and internationally. She acts as an Editorial Advisory Board for Social Responsibility Journal published by Emerald Group Publishing Limited. She also acts as Members of Board in Issues in Social and Environmental Accounting (Issues in SEA), Advisory Board of Emerald Book Series on Developments in Corporate Governance and Responsibility and Editorial Board Member of Journal of Corporate Responsibility and Governance. She also has been awarded "Outstanding Author Contribution Award Winner" by Emerald Literati Network Awards for Excellence 2012. In the year 2016, she has been awarded "Diamond Award" for the idea and product innovation by The 3rd International Innovation, Design and Articulation (i-idea 2016).

Corina Joseph An Associated Professor in Accounting graduated from Curtin University, Australia. Her PhD thesis is on 'Sustainability Reporting on Malaysian Local Authority Websites'. Corina's research interest is in Public Sector Accounting and Governance, Corporate Governance, Sustainability and CSR Reporting and Integrity Framework. She publishes several articles in several international referred journals. She serves as Editorial Board Members at several international journals. She is now working at the Faculty of Accountancy, Universiti Teknologi MARA, Sarawak, Branch, Malaysia.

Index

A
Assurance auditing, 64
Assurance report, 42, 49, 51
Assurance service, 49, 50
Assurance service contract, 51

B
Basis for license audit, 47
Basis for service audit, 48

C
Conflicts of interest, 3–6, 9–20, 107, 151, 153, 239
Corporate influence, 3–5, 9
Customer satisfaction, 168

F
Food and Drug Administration (FDA), 14, 15, 17, 30, 34, 36, 37, 87, 97, 98, 111, 165, 183, 191

H
Health sector, 27, 74, 218, 221, 230, 231

I
ISAE 3402, 41, 42, 49–52, 56, 58–62

J
JCI audit, 48

L
Leveling and registration audit, 48

N
National Institutes of Health, 5, 13, 15, 16, 18

O
Open Payments, 105–107, 111
Outsourcing, 41–46, 48–50, 55, 60, 163, 176, 177, 180, 188

P
Physician Payments Sunshine Act, 17
Private healthcare organization, 41, 42, 47–52, 56, 61, 62
Private Hospital Regulation, 47

R
Regularity audit, 47
Request for assurance, 49, 56
Research funding, 6
Research misconduct, 120, 134, 152

S
Science ethics, 106, 112, 115, 116, 143, 150, 152
Scientific integrity, 10, 12, 14, 18
Service organization, 42, 49
Support services, 41–47, 49–57, 60, 61

K. T. Çalıyurt (ed.), *Integrity, Transparency and Corruption in Healthcare & Research on Health, Volume I*, Accounting, Finance, Sustainability, Governance & Fraud: Theory and Application, https://doi.org/10.1007/978-981-15-1424-1

T

Tobacco, 3, 4, 6–9, 12, 18, 20, 152, 155, 171, 175, 198

Transparency, 4, 9, 16, 29, 65, 66, 69, 71, 74, 87, 91, 92, 95, 98, 105–107, 117, 118, 121, 125–133, 135, 136, 182, 188, 219, 238–240

U

User entity, 49

The manufacturer's authorised representative in the EU is Springer Nature Customer Service Centre GmbH, Europaplatz 3, 69115 Heidelberg, Germany. If you have any concerns regarding our products, please contact ProductSafety@springernature.com

Printed and bound by CPI Group (UK) Ltd, Croydon, CR0 4YY
25/09/2026
02222898-0008